Animal Eyes

Oxford Animal Biology Series

Editors

Professor Pat Willmer is in the School of Biology at the University of St Andrews.
Dr David Norman is Director of the Sedgwick Museum at the University of Cambridge.

Advisers

Mark Elgar (Melbourne)	Gideon Louw (Calgary)
Charles Ellington (Cambridge)	R. McNeill Alexander (Leeds)
William Foster (Cambridge)	Peter Olive (Newcastle)
Craig Franklin (Queensland)	Paul Schmid-Hempel (Zurich)
Peter Holland (Reading)	Steve Stearns (Yale)
Joel Kingsolver (North Carolina)	Catherine Toft (Davis)

The role of the advisers is to provide an international panel to help suggest titles and authors, to ensure individual countries' teaching needs are met, and to act as referees.

The aim of the Oxford Animal Biology Series is to publish attractive supplementary texts in comparative animal biology for undergraduates studying biological science by adopting a fresh, integrated approach. The series has two distinguishing features. One, that topics within each book will be addressed using examples from throughout the animal kingdom, looking for parallels that transcend taxonomy; and two, all aspects of the topics will be chosen to match existing and proposed courses and syllabuses, while taking into account the depth of coverage needed and the amount of space available. Further reading sections, consisting mainly of review articles and books, will guide the student into the literature at greater depth. The series will be international in scope, both in species used as examples and in references to scientific work.

Animal Eyes

Michael F. Land

Professor of Neurobiology, University of Sussex, UK

AND

Dan-Eric Nilsson

Professor of Zoology, University of Lund, Sweden

UNIVERSITY PRESS

OXFORD
UNIVERSITY PRESS

Great Clarendon Street, Oxford OX2 6DP

Oxford University Press is a department of the University of Oxford.
It furthers the University's objective of excellence in research, scholarship,
and education by publishing worldwide in

Oxford New York

Auckland Cape Town Dar es Salaam Hong Kong Karachi
Kuala Lumpur Madrid Melbourne Mexico City Nairobi
New Delhi Shanghai Taipei Toronto

With offices in

Argentina Austria Brazil Chile Czech Republic France Greece
Guatemala Hungary Italy Japan South Korea Poland Portugal
Singapore Switzerland Thailand Turkey Ukraine Vietnam

Published in the United States
by Oxford University Press Inc., New York

© Oxford University Press, 2002

The moral rights of the author have been asserted
Database right Oxford University Press (maker)

First published 2002
Reprinted 2004 (with corrections), 2005, 2006, 2008

British Library Cataloguing in Publication Data
Data available

Library of Congress Cataloging in Publication Data
Land, Michael F.
  Animal eyes/Michael F. Land, Dan-Eric Nilsson.
    p. cm.—(Oxford animal biology series)
  Includes bibliographical references (p. 00).
  1. Eye. I. Nilsson, Dan-Eric. II. Title. III. Series.
QL949.L26 2001   573.8′8—dc21   2001036500

ISBN 978 0 19 857564 1 (hbk)
    978 0 19 850968 4 (pbk)

Typeset by EXPO Holdings, Malaysia
Printed in Great Britain
on acid-free paper by Biddles Ltd., King's Lynn, Norfolk

For Rosemary and Maria

Preface

'The eye' to most people means an eye like ours, a single chambered camera-like structure with a retina in place of the film, or the CCD array. Most know, too, that insects have compound eyes with many lenses, but how many people can answer the question: does the insect see the multitude of images beloved of Hollywood horror films, or a single image similar to our own? We use this example to point out that, even to most biologists, eyes remote from our own are poorly understood and come in only one or two varieties. This hugely underestimates the diversity of eye types: there are at least ten quite distinct ways that eyes form images. Some of these such as pin-holes and lenses are familar, but others are more exotic. These include concave mirrors, and arrays of lenses, telescopes and corner reflectors. Some have been known about for centuries (the first demonstration of the inverted image in a mammalian eye was in 1619) but a number are discoveries of the last few decades and have yet to find their way into textbooks of either biology or optics. Some of these eye-types have counterparts in optical technology, but by no means all. Some are still finding applications: for example, the mirror-based optical system of the compound eyes of shrimps and lobsters has recently found a use as the optical basis of wide angle X-ray lenses.

It is our aim in this book to provide a comprehensive account of all known types of eye. We take the diversity of optical mechanisms as a framework, but many other aspects of the structure and function of eyes are also dealt with. Visual ecology – the ways that eyes are specifically adapted to the lifestyles of the animals that bear them – is another important theme. As humans we tend to think of vision as a general-purpose sense, supplying us with any kind of information we require. For most other animals this is not so. Predators and prey, for example, have different visual requirements: foxes and rabbits have different eyes and different visual systems, as have dragonflies and mosquitoes. Similarly, a sedentary clam lives in a different world from a flying insect, and the optical requirements are quite different.

Behind the diversity of eye types is the majesty of the evolutionary process, and this is where we begin the book. The origins of eyes, and the ways in which they reached their present highly developed states has posed an intriguing series of problems from Darwin onwards. The debates still rumble on, particularly about the early origins of eyes before the great Cambrian radiation event gave us most of the eye types we see in animals today. Chapter 1 addresses these questions, and provides a context in which eyes can be seen as different solutions to problems that are, in many respects, similar.

As well as diversity, we are concerned with the 'design philosophy' of eyes. What are the physical constraints on the way an eye performs its functions, and how are these addressed by the different types of eye? To answer this it is necessary first to have some information about the properties of light that are of importance for vision, and this we provide in Chapter 2. We are then able to explore the ways that eyes achieve important aspects of their function, such as good spatial resolution, and (especially for animals that live in dim environments) adequate sensitivity. This is the purpose of Chapter 3, which is devoted to the question 'What makes a good eye?'. This in turn provides a background for assessing the capabilities of the panoply of different eye types, presented in the subsequent five chapters. The ninth and final chapter examines another aspect of the way eyes are used: their movements. Eyes sample the world not only in space but in time, and the movements that they make are as important a part of the process of extracting information as are the optical systems that provide them with spatial resolution.

The book is not aimed at any one readership. It will be of value to undergraduates in Biology and Neuroscience programmes, and to anyone engaged in the study of vision at the post-graduate level. Students and practitioners of ophthalmology and optometry will find it interesting as a background to the study of the human eye, and optical physicists and engineers will find that nature has come up with solutions that they will not have met before.

Aware that many biologists will want the story without too much mathematical detail, we have used Boxes for some of the more complex sections. Equally, however, serious students will want to make use of some of these sections as they contain important 'how to do it' information. For example, Box 5.1 shows how to find the focal length and image position in any optical system of reasonable complexity. We have not provided a complete bibliography justifying every statement in the book, but given references to reviews where the original literature can be found, and to key works, with a bias towards the more recent.

We would like readers to enjoy the book, and share in our enthusiasm for the beauty, intricacy and the logic of animal eyes that has kept us intrigued, and busy, for a total of 60 years.

Acknowledgments

We are especially grateful to Cathy Kennedy, our editor at Oxford University Press for most of the time we were working on the book. She read everything, provided us with the intelligent layman's feedback we needed, and managed to cajole and enthuse in just the right mixture. We thank all those people who kindly read and commented on sections of the book, and all those who supplied us with original illustrations. Their names appear in the figure captions.

We thank the following publishers for allowing us to use copyright figures: Academic Press (Figs 5.13, 6.10d&f), The Association for Research in Vision and Ophthalmology (6.10e), Cambridge University Press (4.1, 4.3), Kluwer Academic Publishers (3.8), Macmillan Magazines Ltd. (5.12), Sigma xi (1.1a) and Springer-Verlag (2.7a, 5.13, 9.8, 9.13). The authors are referred to in the figure captions, and full citations appear in the reference list.

Contents

1 The origin of vision 1

The first eyes 1
What is an eye? 4
Why and from what did eyes evolve? 6
The course and pace of eye evolution 7
Did eyes evolve once or many times 10
Summary 15

2 Light and vision 16

The nature of light 16
Light intensity 19
Contrast 24
Wavelength and colour 24
Polarization 29
Summary 32

3 What makes a good eye? 33

Fundamentals 33
Resolution 36
Sensitivity 47
Conclusions 53
Summary 55

4 Aquatic eyes: the evolution of the lens 56

Evolutionary origins 56
The pin-hole eye of Nautilus 57
Spherical lenses 59
Lenses with refractive index gradients 60
Eyes of fish and cephalopods 63
Matching eye to environment 66
Eyes with non-spherical lenses 69
Summary 71

5 Lens eyes on land 72

A new optical surface 72
Basic optics of cornea and lens 73
Variations on the lens/cornea theme in land vertebrates 81
Amphibious eyes 93
Invertebrate eyes with corneal optics 94
Summary 102

6 Mirrors in animals 104

Mirrors in eyes 104
The physical optics of animal reflectors 114
Uses of multilayer reflectors in structures other than eyes 118
Summary 124

7 Apposition compound eyes 125

Origins 125
A little history: apposition and neural superposition 127
Basic optics 131
Ecological variations in apposition design 142
Summary 155

8 Superposition eyes 156

Introduction – the nature of superposition imagery 156
Refracting superposition 159
Superposition and afocal apposition 168
Reflecting superposition 172
Parabolic superposition 176
Summary 177

9 Movements of the eyes 178

How humans acquire visual information 179
Are other animals like us? 183
Insect flight behaviours seen as eye movement 186
Why not let eyes wander? Some consequences of image motion 187
Exceptions: rotational scanning by one-dimensional retinae 193
Summary 200

Principal symbols used in the text 201

References 202

Index 215

1 | The origin of vision

The first eyes

When the first life appeared on earth, the sun shone as brightly as it does today, and a human eye would have been able to see as well then as it does now. Yet, the first forms of life were unable to see the sun or the early landscape basking in its light. Although life has existed for several billion years, animals advanced enough to make use of vision have only been around for little more than half a billion years. If we trace eyes back through the fossil record, it appears that proper vision originated in the early Cambrian, some 530 million years ago. The Cambrian animals were not the same species as exist today, but nearly all the modern phyla had rapidly come into existence, fully equipped with eyes as far as can be told from the fossils. Only 20 million years earlier, towards the end of the Precambrian, the forms of life seem to have been much simpler. The Precambrian fossil fauna totally lacks large mobile animals that could benefit from good vision. It is even hard to identify any animals at all in the fossil remains of Precambrian organisms. But something remarkable seems to have happened in the interface between the Precambrian and the Cambrian. Within less than five million years, a rich fauna of macroscopic animals evolved, and many of them had large eyes. This important evolutionary event is known as the Cambrian explosion.

The Cambrian fossils have been gradually deciphered since 1909 when the paleontologist Charles Walcott started to analyse the 515-million-year-old rock of the Burgess shale in Canada. What Walcott found was the well preserved remains of a marine fauna, presumably from shallow water. The fauna was dominated by arthropods of many different types, but also contained representatives of numerous other phyla. Some of the interpretations of the Burgess shale fossils indicated the appearance of many enigmatic types of animals which did not seem to belong to any phyla remaining today. Subsequent and more careful analyses have demonstrated that nearly all of the Cambrian animals were indeed early representatives of modern animal phyla (Conway-Morris 1998). After the discovery of the Burgess shale fauna, two even better preserved fossilized faunas have been found: the Chengjiang fauna of central China, and the Sirius Passet fauna of northern Greenland. These two faunas are particularly interesting because they are older than the Burgess shale and offer a close look at animal life very soon after the Cambrian explosion. Amazingly, these earlier faunas were not very different from that preserved in the Burgess shale. It thus seems that essentially modern types of animals, many with large eyes, evolved within a few million years from ancestors that for some reason were not large or rigid enough to leave much fossil traces.

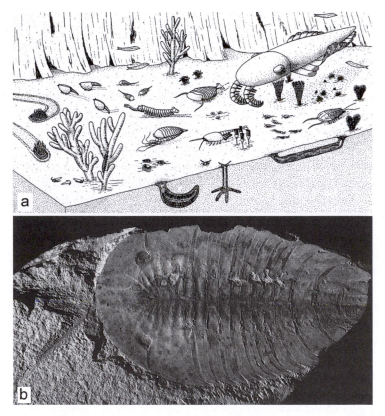

Fig. 1.1 Evidence of the first real eyes comes from Cambrian fossils. The first faunas with mobile animals originated at the onset of the Cambrian era, some 540 million years ago, during a rapid evolutionary event called the Cambrian explosion. In the course of a few million years, bilaterally symmetric, macroscopic and mobile animals evolved from ancestors that were too small or soft bodied to be preserved as fossils. The product of the Cambrian explosion was not just a few species, but an entire fauna (a) including nearly all the animal phyla we know today. The invention of visually guided predation may have been the trigger for this unsurpassed evolutionary event. Among the very first Cambrian animals, numerous species had prominent eyes. An early example is the arthropod *Xandarella* (b) from Chengjiang, China. Unfortunately, fossils generally reveal very little, if anything about the type or structure of these ancient eyes. (a) from Briggs (1991), originally adapted from Conway Morris and Whitington (1985), (b) from Xianguang and Bergström (1997), with the authors' permission.

In the early Cambrian faunas, trilobites and other arthropods were abundant, and it is reasonable to believe that they viewed the world through compound eyes. In trilobite fossils it is often possible to see the facets of the compound eyes, but in other Cambrian fossils, the eyes are just visible as dark imprints with no detailed structures preserved. Figure 1.1 shows a Cambrian fossil and reconstructed creatures with prominent eyes. From the abundance of eye-bearing species, and from the sizes of their eyes, it seems that vision was no less important in the early Cambrian than it is today. The fossils clearly tell us that, from their first appearance, macroscopic mobile animals were equipped with eyes (Fig. 7.21).

Vertebrate eyes cannot be traced back to their very beginning in the same way that arthropod compound eyes can. The first Cambrian faunas did contain animals that are believed to be early chordates (*Pikaia*, Fig. 1.2) but like the modern lancelets (*Amphioxus*), it appears that it did not have proper eyes. Chordates with obvious eyes appear first about 25 million years after the Cambrian explosion. An early Ordovician conodont chordate (Fig. 1.2) had such large eyes that it must have had better vision than most other animals of its time. Some 30 million years after the conodonts the first fish appeared, and the chordate lineage became a much more dominant component of the ecological community. Eye evolution is thus largely a story about what happened in the Cambrian, and thereafter it was only the colonization of land that led to general evolutionary events in vision.

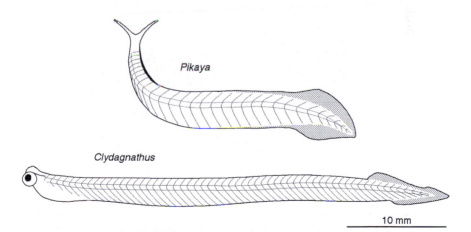

Fig. 1.2 Vertebrate eyes cannot be traced all the way back to the Cambrian explosion. Although chordates, such as *Pikaya*, were present in the early Cambrian, they did not have any obvious eyes. The first evidence of eyes in this phylum are found in conodont animals which appeared 30 million years later. One example is *Clydagnathus* which had unusually large eyes of the single chambered type, possibly similar to present day vertebrate eyes. *Pikaya* redrawn after Gould (1989), *Clydagnathus* redrawn after Purnell (1995).

What is an eye?

Thinking of modern animals, it is easy to take eyes for granted, being as natural and integral parts of a typical animal body as a mouth and a gut. In reality, there are numerous animal species that have no eyes at all, and still manage well enough to be extremely abundant. Among the 30 or so different animal phyla, only about a third contain species with proper eyes, another third has small light-sensitive organs, but no real eyes, and the final third has no obvious specializations at all for detecting light.

Before we continue into questions of how and why eyes evolved, it is useful to work out a definition of what an eye is. A survey of the animal kingdom does not give any obvious guidance here because nearly every imaginable intermediate exists between the acute vision of an eagle and the simple light sensitivity of an earthworm. However, there is an important distinction between these two extremes: both are able to measure the amount of light, but only the eagle can compare the amount of light coming from different directions. Even the simple pit eye of a planarian flatworm (Fig. 1.3) has some ability to compare intensities in different directions, albeit with much less precision than an eagle's eye. What matters is the ability to detect an image, no matter how crude it is. We term this visual modality *spatial vision*.

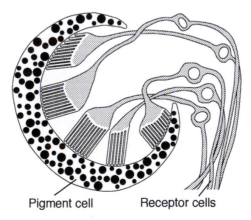

Pigment cell   Receptor cells

Fig. 1.3 Animals that are small and not very swift often make do with eyes of uncomplicated design. The eyes of turbellarian worms include many such examples. Here in the eye of *Bdellocephala brunnea* a small number of receptor cells share a pigment cup. The receptive segments (microvilli assembled into rhabdoms) occupy different positions in the pigment cup and thus receive light from different directions. This allows the animal to simultaneously discriminate between the brightness of different parts of a very crude image. Even though the amount of pictorial (spatial) information is minuscule compared to that of a human eye, the flatworm eye is clearly an organ capable of spatial vision. In terms of function there is only a gradual difference between the eyes of flatworms and humans. Redrawn from Kuchiiwa et al. (1991).

In principle, the simplest way to produce spatial vision would be to have two light sensitive cells shielded so that they do not pick up light from exactly the same direction. This would qualify as spatial vision if the nervous system of the animal were able to make a comparison between the responses of the two cells. From such a rudimentary kind of image vision it is possible to envisage a gradual refinement, by adding more light sensitive cells, each with is own unique angular sensitivity. (In practice no known eyes have only two receptors, but there are many with numbers as small as ten). The analogy with a printed image comes naturally: the more numerous and densely packed the dots that make up the picture are, the more spatial (pictorial) information is contained in the picture. On qualitative grounds we would thus group together the pit eye of the planarian with the eagle's eye, as organs of spatial vision, but exclude the earthworm's light sensitivity. Just for the sake of the discussion in this chapter, we shall define real eyes as organs for spatial vision. The evolution of eyes then becomes the evolution of spatial vision.

Even though our definition is clear-cut, there are examples of simple visual organs which need some further comment. Large numbers of small invertebrates have eye spots consisting of a few light sensitive cells (photoreceptor cells) shielded from one side by dark pigment (Fig. 1.4). Such an arrangement will make the photoreceptor cells directional, but it does not imply that there is spatial vision. It is perfectly possible, and probably useful to many invertebrates, to make directional light intensity readings without being able to simultaneously

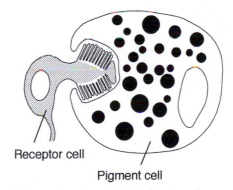

Receptor cell

Pigment cell

Fig. 1.4 The simplest visual organs, here exemplified by the eye spot of a larval trematode worm, *Multicotyle purvisi*, consist of a single photoreceptor cell shielded on one side by screening pigment. This organ provides no spatial information in the sense that light from different directions can be simultaneously discriminated, but by comparing the signal from two such organs or by moving the body, the bearer can navigate towards brighter or darker places, or keep a certain body orientation. If eyes are defined as organs capable of spatial vision, the depicted eye spot would not qualify as a true eye, but it is certainly an evolutionary forerunner to real eyes. The next level of functional complexity would be an eye such as that of Fig. 1.3. Redrawn from Rhode and Watson (1991).

compare the intensity in different directions. Directional photoreceptors may serve important tasks such as guiding phototaxis and informing about body orientation. A well-known example occurs in the larvae of flies, where a small group of receptors shielded from behind makes it possible for the larvae to steer away from a light source: they simply turn their heads from side to side, and if the current movement causes an increase in light intensity they turn further in the opposite direction (Fraenkel and Gunn, 1961, Chapter VI). Undirectional photoreceptors may also be sufficient for other tasks such as knowing the time of day or depth in the sea. We are now in the position to use our definitions in the discussion of eye evolution.

Why and from what did eyes evolve?

As we have seen, the fossil evidence suggests that a large range of visually guided animals evolved in a very short time during the early Cambrian. Did their eyes evolve from scratch at that time, or might their ancestors already have had some precursor of real eyes. The fossils do not give a clear answer here, but they provide some interesting clues. Fossils formed towards the end of the Precambrian reveal tracks made in the seafloor, and these increase in frequency as the Cambrian explosion approaches. From the size and appearance of these tracks it seems that they were made by small (a few millimetres) worm-like animals slowly crawling on the surface of the seafloor. The fact that the actual animals are not fossilized may indicate that they were soft-bodied creatures. If they belong to the ancestors of the early Cambrian faunas they would have had to increase considerably in size at the initial phase of the Cambrian explosion. Skeletons and rigid protection seem to have evolved along with the larger bodies. The first evidence of such structures comes at the very end of the Precambrian. Fossils of small shells, or fragments of shells, typically in the size range of 2–10 mm, known as the 'small shelly fauna', very closely preceded the Cambrian explosion.

Lacking direct evidence, we may try to deduce what types of visual organs, if any, the late Precambrian animals had. If the animals were small and slow moving they would have benefited primarily from directional photoreceptors. As we shall see in Chapter 3, the physical nature of light makes the potential for spatial vision increase with eye size. Visual organs that are very small cannot resolve much spatial detail, no matter how well they are designed. The small and slow animals of the Precambrian would thus have had neither the abilities nor the reason to acquire proper eyes with spatial vision. The first Cambrian faunas, on the other hand, included a wealth of large and mobile creatures with large eyes and presumably good spatial vision. It is tempting to speculate that a few species of late Precambrian animals became large enough to acquire good spatial vision and improved mobility, and became the first visually guided predators.

Such an ecological invention would have put a tremendous selection pressure on a large part of the fauna, and forced other species to evolve protective measures such as body armour or shells, avoiding exposure by deep burrowing, or developing good vision and mobility themselves. These possibilities are indeed the key characteristics of the early Cambrian faunas, supporting the idea that the introduction of visually guided predation altered much of the ecological system and fuelled the Cambrian explosion. Because both vision and speed of locomotion can improve by a general increase in size, visually guided predation offers an understanding of the very sudden appearance of macroscopic animals. In this scenario the small shelly fauna may have been the very first stages of an arms race between predators and prey, where rigid structures for protection and mobility evolved along with the first proper eyes.

The course and pace of eye evolution

To understand eye evolution is partly to understand how spatial vision can arise. Whatever the role of vision was in the Cambrian explosion, it seems reasonable to assume that proper eyes were preceded by directional photoreceptors. Already the simplest attempts to achieve directionality requires the employment of some sort of optical mechanism, and the evolution of spatial vision can be seen as an evolution of such mechanisms. An organ for directional vision must, in addition to one or more photoreceptor cells, contain some structure which provides the directionality. In its simplest form this is achieved by a shield of dark pigment (Fig. 1.4). There are two fundamentally different ways by which spatial vision can evolve from a shielded photoreceptor: Either more photoreceptors are added to exploit the same pigment shield, or the visual organ is multiplied in its entirety. The two alternatives lead to simple (single-chambered) and compound eyes respectively. In Fig. 1.5 the primitive eye of a clam illustrates a case which would probably turn into a compound eye if vision were to improve any further. During the early stages of eye evolution there would be little difference between the efficiency of the two solutions – single-chambered or compound. It is only later, when visual performance is maximized for a constrained eye size, that the simple eye will turn out to be a better solution (the relative merits of the different types of eye will be explained in Chapter 3).

Irrespective of whether evolution originally takes the path towards a simple or a compound eye, shielding will soon turn out to be an inefficient mechanism on its own. As the spatial resolution is improved by adding more picture elements, the directionality of each photoreceptor will need to improve as well. It is at this stage in eye evolution where more elaborate optics, in the form of lenses or mirrors, will significantly improve the design. Because even the slightest degree of focusing is better than none at all, lenses or mirrors can be introduced gradually, with a continuous improvement in performance.

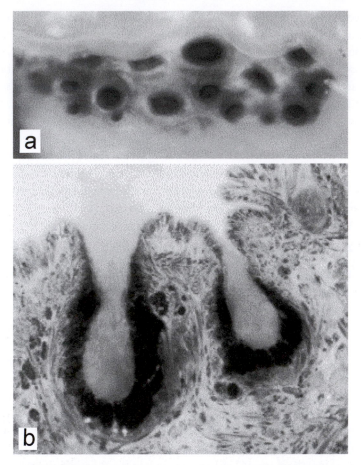

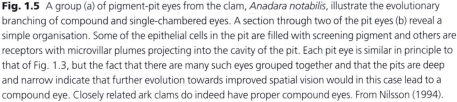

Fig. 1.5 A group (a) of pigment-pit eyes from the clam, *Anadara notabilis*, illustrate the evolutionary branching of compound and single-chambered eyes. A section through two of the pit eyes (b) reveal a simple organisation. Some of the epithelial cells in the pit are filled with screening pigment and others are receptors with microvillar plumes projecting into the cavity of the pit. Each pit eye is similar in principle to that of Fig. 1.3, but the fact that there are many such eyes grouped together and that the pits are deep and narrow indicate that further evolution towards improved spatial vision would in this case lead to a compound eye. Closely related ark clams do indeed have proper compound eyes. From Nilsson (1994).

It may seem that the evolution of an eye would be both difficult and time-consuming, and it has frequently been argued that a great deal of good fortune would be required for eyes to evolve. But the truth is that eyes can evolve gradually from the simplest form of light sensitivity to a perfectly focused eye with all its intricacies. The only external factor needed is an ongoing selection favouring better spatial resolution. The time required for an eye to evolve then becomes a question of how much and how fast the tissue structures must change. Using a theoretical model, Nilsson and Pelger (1994) calculated that a light-sensitive patch on the skin can gradually turn into a typical vertebrate or octopus eye by some 2000 sequential changes of 1 per cent in length, width, or protein density of the cells in the skin patch. Throughout this sequence, spatial resolution improves continuously, provided selection favours it. It may sound remarkable that only 2000 of such tiny modifications can lead all the way to an advanced eye, but the power of these sequential modifications should not be underestimated: if 2000 sequential modifications of 1 per cent are applied to lengthen a single structure such as a 10-cm-long finger, then it becomes long enough to bridge the Atlantic ocean. In eye evolution there is a multitude of different structures that are modified in various ways, but the important point is that the amount of required structural change can be quantified, and thus used to estimate the time it would take. Even with very slow evolution where the modification from one generation to the next is as little as 0.005 per cent, Nilsson and Pelger concluded that an eye may evolve in less than 400 000 generations. If each generation is completed within a year, it means that an eye can evolve in less than half a million years. This calculation would have been a good cure for the famous 'cold shudder' that

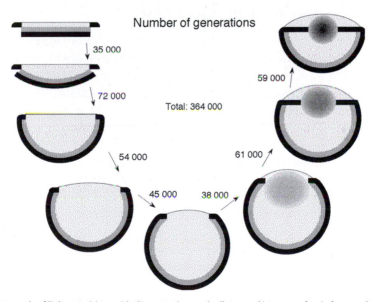

Fig. 1.6 A patch of light sensitive epithelium can be gradually turned into a perfectly focussed camera-type eye if there is a continuous selection for improved spatial vision. A theoretical model based on conservative assumptions about selection pressure and the amount of variation in natural populations suggest that the whole sequence can be accomplished amazingly fast, in less than 400 000 generations. The number of generations is also given between each of the consecutive intermediates that are drawn in the figure. The starting point is a flat piece of epithelium with an outer protective layer, an intermediate layer of receptor cells, and a bottom layer of pigment cells. The first half of the sequence is the formation of a pigment cup eye. When this principle cannot be improved any further, a lens gradually evolves. Modified from Nilsson and Pelger (1994).

Darwin felt when he thought about the refined form and function of the human eye: its evolution need not have been as problematic as he feared. Of more importance here is that it allows an understanding of how eyes could evolve so rapidly during the Cambrian explosion.

The actual course of evolution of large eyes with high spatial resolution is reasonably easy to chart, because every step on the way is represented in animal species living today. But this also leads to the question why there appear to be so many intermediates still in existence, when eye evolution can be potentially so fast. The key to this paradox is probably that the apparent intermediates are really end products in the sense that evolution has proceeded to a point where there is no further selection for improved spatial resolution. We have to keep in mind here that visual information is only useful if the animal can improve its behaviour on the basis of it. For every species there is thus a limit to how much spatial information it can use. There is, of course, also a cost involved in making and maintaining eyes, and it is this final balance which determines how much vision each species can afford.

Did eyes evolve once or many times?

A much debated question in eye evolution is how many times eyes have evolved independently. At least some of the controversies on this matter come from a lack of definition of what an eye is. In this chapter we have defined an eye as an organ providing spatial vision. On the basis of the two fundamentally different ways of creating spatial vision, resulting in simple and compound eyes, it seems that eyes must have evolved more than once. Comparing the embryological origin of animal eyes, it is clear that they derive from different tissues in different animal groups (Fig. 1.7). The vertebrate retina develops as an eye cup formed by frontal parts of the brain, but the lens is formed by the skin. In the nearly identical eyes of octopus and squid the lens and the retina both develop from the skin. Even the photoreceptor cells from different phyla display fundamental differences in their basic structure and biochemistry, indicating separate origins dating deep into the Precambrian, long before proper eyes evolved (Salvini-Plawen and Mayr 1977). The photopigment proteins (the opsins) tell a different story because they appear to come from the same origin in all animals. Also the master control genes (Pax-6), which turn on a downstream gene cascade forcing an eye to develop in the embryo, are amazingly similar in all vertebrates and invertebrates investigated so far (Gehring and Ikeo 1999). This may seem to indicate common ancestry, but the problem is that we do not know if the master control gene originally evolved to control the development of an eye with spatial vision, or a directional photoreceptor, or perhaps just the simplest form of light sensitivity (Nilsson 1996). Taking all the evidence together, it seems that the photopigment molecule originated a very long time ago in a common ancestor to all animals, that photo-

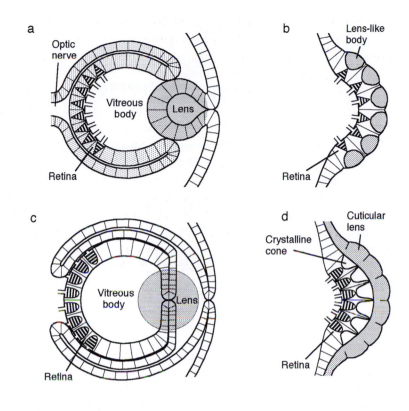

 Ciliary photoreceptor cell   Microvillar photoreceptor cell

Fig. 1.7 The composition of eyes in (a) vertebrates, (b) polychaete fan worms, (c) octopus and squid, (d) insects and crustaceans. Although there are only few ways of making functional eyes, the tissues and morphological components that are recruited vary greatly between animal phyla. The vertebrate retina (a) is produced by the neural epithelium of the brain (light shading) and the lens is formed by an invagination of the epidermal epithelium. In squid and octopus the entire eye is formed as a double epidermal cup, with the bottom of the inner cup being the retina and its fused opening producing the lens. The receptor cells are also fundamentally different in that they contain the visual pigment in either modified cilia (ciliary receptors) or microvilli (rhabdomeric receptors), and in the biochemistry of their transduction machinery. A consequence of the ontogenetic origin of vertebrate eyes is that the receptor axons project towards the vitreous body and have to emerge from the eye through a hole in the retina. The compound eyes of fan worms (b) and arthropods (d) have likewise recruited different types of visual receptor cells, but more importantly they are formed on different parts of the body: as paired structures on the first segment of the head in arthropods and as multiple structures on the feeding tentacles of fan worms. These facts taken together clearly indicate that at least these four cases evolved spatial vision independently, and arrived at two different solutions, the camera eye and the compound eye. Modified from Nilsson (1996).

receptor cells have evolved independently a handful of times and that spatial vision (real eyes) have evolved numerous times. Clearly, there are different levels of homology, and even if the eyes of vertebrates, octopuses and insects did not

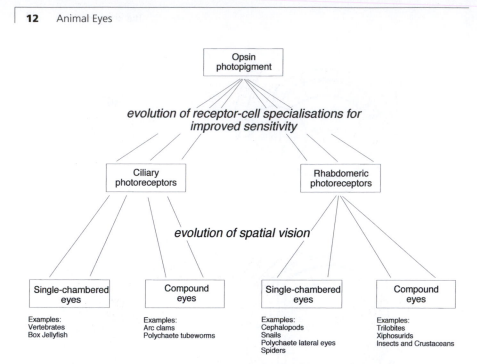

Fig. 1.8 Different levels of homology. All animal eyes share a homologous type of visual pigment but the various specialisations for accommodating large amounts of visual pigment and for transducing the information into electrical signals appear not to be homologous. Instead they are parallel solutions to a common problem of sensitivity and speed of visual receptor cells. The different types of receptor cells have subsequently been recruited independently in a number of instances to produce imaging eyes. Remarkably, there are striking similarities between independently evolved eyes in the chain of control genes that orchestrate eye development. This indicates that the control genes are ancient and were originally associated with opsin transcription and sensory or neural fate in a common ancestor. The levels of homology tell an interesting story about animal evolution. The common ancestor must have been a very simple organism which used light intensity to control its behaviour. Elaborations to obtain some directionality, high sensitivity and response speed may have independently resulted in different receptor types and visual organs such as those shown in Fig. 1.4. This evolutionary process must have been mostly finished in the Precambrian, and the results were used by animals which had more complex responses to light. The evolution of macroscopic and swiftly mobile animals in the early Cambrian led to a strong selection for improved spatial vision which produced the first proper eyes with imaging optics. This was a parallel process in several animal phyla, and the superficial similarities between separate evolutionary lines is due to design constraints discussed in detail in following chapters. Eyes of different phyla can thus be said to be both homologous and non-homologous depending on the level of organisation considered.

originate in a common ancestor, insects and octopuses may have homologous photoreceptor cells, and all three may share a homologous visual pigment (Fig. 1.8).

If eyes had not evolved, life on earth would have come out very differently. More than any other organs, eyes have shaped the evolution of animals and ecosystems since the Cambrian explosion. The result is an enormous range of

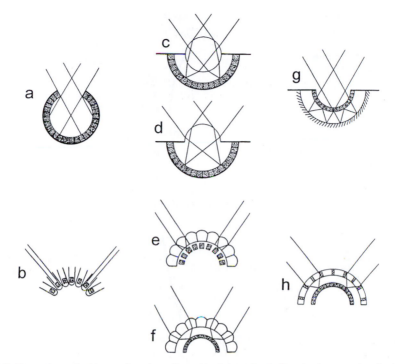

Fig. 1.9 The major optical types of eye found in multicellular animals, forming images using shadow (a,b), refracting devices (c-f), and reflectors (g,h). The receptors are shown in stipple. Relevant chapters are designated in bold type. (a) Pit eye, forerunner of other single-chambered eyes (c,d,g). Found in planarians, many annelids and molluscs and in advance pinhole form in *Nautilus* **4** (b) Basic compound eye, in which each receptor is shielded from its neighbour by a simple pigment tube. Found now only in a few bivalve molluscs **7**. (c) Aquatic lens eye, found in fish and cephalopod molluscs **4**. (d) Corneal lens eye found in terrestrial vertebrates and spiders and some insect larvae **5**. (e) Apposition compound eye, characteristic of diurnal insects and crustaceans (e.g. bees and crabs). Each receptor cluster has its own lens **7**. (f) Refracting superposition compound eye found in animals from dim environments (e.g. moths and krill). Here many lenses contribute to the image at each point on the retina **8**. (g) Single chambered eye where the image is produced by a concave mirror. The best example is in the scallop *Pecten* **6**. (h) Reflecting superposition eye. Similar to (f) except that the lenses are replaced by mirrors. Found in decapod shrimps and lobsters **8**. Modified from Land (1981). An evolutionary tree of the animal kingdom can be found in Fig. 1.10.

eye types using pin-holes, lenses, mirrors, and scanning devices in various combinations to acquire information about the surrounding world (Fig. 1.9). Even though vertebrate eyes are the ones that first come to mind, they are poor examples in the sense that they are nearly all alike. Invertebrates, in particular the arthropods, molluscs, and annelids display an inventiveness which is mind-boggling. In each of these three phyla both camera type eyes and compound eyes have evolved independently. Not all eyes are paired and placed on the head: there are chitons with eyes spread over their dorsal shell, tube worms with eyes

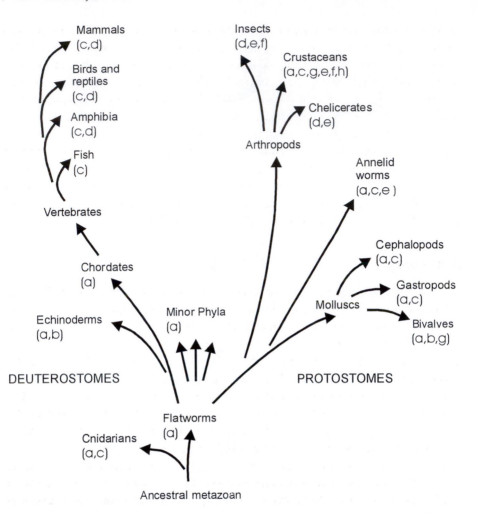

Fig. 1.10 Diagram of the evolutionary relationships of the major animal groups. At the base of the tree are the cnidarians (hydra, jellyfish) with only two tissue layers, then the flatworms with three layers but no coelom (body cavity). All the later phyla have a body cavity. There are two main branches, the deuterostomes which led to the starfish and the vertebrates, and the protostomes which gave rise to the molluscs, annelids and the arthropods. The division is based largely on embryological differences. Echinoderms, chordates, molluscs, annelids and arthropods are all phyla – the major divisions of the animal kingdom. Vertebrates and the three arthropod groups are subphyla, and the four vertebrate groups are classes, as are the three mollusc groups. Recent molecular evidence links the arthropods with other moulting animals, including the nematode worms, and distances them somewhat from the annelids which were formerly thought to be close to the ancestral arthropod line.

The main types of eye present in each group are indicated by letters, which refer to the diagram in Fig. 1.9. In general the protostomes have microvillous receptors (Fig. 1.8) and the deuterostomes have ciliary receptors, but there are numerous exceptions.

on their feeding tentacles and clams with eyes on the mantle edge. Even more surprising is the occurrence of eyes in creatures without a brain: box jellyfish have camera type eyes feeding information into a simple ring-shaped nervous system, and some dinoflagellates (unicellular algae) have a lens and a retina-like structure all in the one cell (Greuet, 1982). Eyes can be less than a tenth of a millimetre, as in some water fleas, and up to 300 mm in giant squid and the ichthyosaurs (extinct marine reptiles). This enormous range of sizes, designs and placement of eyes reflect the versatility of vision, and it gives clear indication that eyes can evolve easily, recruiting whatever tissue is at hand, and become superbly optimized for the lifestyle of the bearer. In the remaining chapters of this book we work our way through the fundamentals of eye design and explain the function and rationale of all the different types of eye.

Summary

1 Most of the types of eye that we recognize today arose in a brief period during the Cambrian, about 530 million years ago. The development of better eyes coincided with increases in size, speed and armour, as predation became a common way of life.
2 We define an eye as an organ of spatial vision, in which different photo-receptors view slightly different directions in space. Such an eye may achieve this simply by shadow (planarian eye) or by more sophisticated optical arrangements. This definition would exclude structures, such as the photo-receptors of fly larvae, where the shadowing is unidirectional. Spatial vision requires simultaneous comparison of light levels in different directions.
3 The evolution of advanced eyes need not have taken huge periods of geological time. It has been estimated that evolution from a patch of shadowed photosensitive tissue to an eye resembling that of a fish could have taken as little as half a million years.
4 The evidence strongly supports the conclusion that modern types of eye evolved independently many times. However, photoreceptor cells fall into a much smaller number of types, and so probably predate eyes themselves. The photopigment rhodopsin appears to have been present in a common metazoan ancestor, and certain genes controlling eye development also had very early origins. Thus there are different levels of homology in eye evolution.
5 We preview the immense variety of eye types, especially amongst inverte-brates, before describing them in detail later in the book.

2 | Light and vision

Eyes are devices for extracting useful information from the light reflected or emitted from objects in the world around us. Most of this book is devoted to a detailed account of how this is done, but before embarking on that saga we need briefly to explore some of the properties of light that are important for vision.

Light usually travels in straight lines with little loss in air or clear water. For an advanced eye with good resolution this means that the geometrical features of an object can be represented in the pattern on the retina, and also that the relative locations of different objects in the world can be determined. Light thus supplies most of the information needed to work out both *where* an object is and *what* it is. In addition to geometric information, light provides other cues to the identity of objects. Light interacts with matter in many different ways. It can be reflected, transmitted, absorbed, or scattered, and all these transformations depend on wavelength. This in turn means that most light is coloured, when seen by an animal with the facility to detect these spectral differences. Some animals, though not ourselves, make use of another physical property of light – polarization – to work out the direction of the sun, and to detect reflecting surfaces.

In this chapter we consider first what sort of energy light is, and what cues it provides for vision; second, how much light is available in the environment and how this is measured; and finally how the photoreceptors in the eye capture light and signal its more subtle properties, such as wavelength distribution and polarization structure.

The nature of light

It has never been easy to understand how light works. Isaac Newton (1642–1727) thought that light was a stream of 'corpuscles' whose trajectories are what we think of as rays. Rays (lines that are straight in a vacuum but which can be reflected by mirrors and refracted by prisms and lenses) provide a very simple and convenient way of describing how images are formed, so long as the structures that bend the rays are large compared with the wavelength of light, which is about 0.5 μm. Some phenomena, however, are not well described by ray optics. Interference effects, such as the colours of bubbles and oil films, and 'Newton's rings' (the circular patterns made when a convex lens contacts a plane glass block) can only be understood in terms of the interactions of waves. Newton's contemporary Christian Huygens first formulated the wave theory in a form that could also take into account the ray-like behaviour of light (Fig. 2.1a). However,

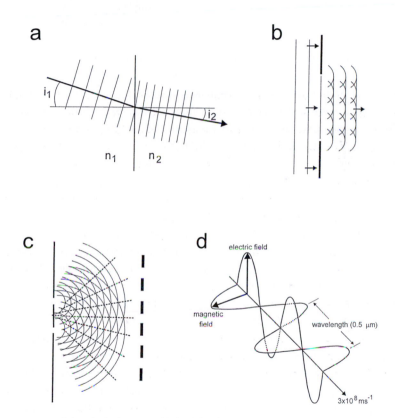

Fig. 2.1 Aspects of the physical nature of light. (a) Refraction can be thought of as the bending of a ray (thick line), or as the slowing down of a series of wavefronts (thin lines) as they enter a higher refractive index medium. This slowing bends the wavefront, resulting in Snell's law ($n_1\sin i_1 = n_2\sin i_2$). Rays are perpendicular to wavefronts. (b) Wavefronts passing through an aperture. In the Huygens–Fresnel scheme each point on the wavefront is an emitter of secondary wavelets. These add in the direction of travel and cancel in other directions so that the plane wavefront is retained, but at the edges of the aperture light spreads laterally, resulting in diffraction. (c) Interference produced by Young's slits. Light from a single source passing through two narrow slits interferes to produce a pattern where wavefronts are in phase and add (*dotted lines*) or are out of phase and cancel. This results in a pattern of light and dark stripes. (d) Propagation of light according to Maxwell. Light consists of oscillating electric and magnetic fields perpendicular to each other. Each element (photon) has a fixed electric field (E-vector) direction, and a fixed wavelength, and propagates through space at a fixed velocity.

the authority of Newton was such, even beyond the grave, that the wave theory made little progress in the eighteenth century. It was Thomas Young's demonstration in the early 1800s that light passing through two narrow slits produces an interference pattern that revived the wave theory and gave it experimental solidity (Fig. 2.1c). The interference of sea waves passing through gaps in breakwaters provides a helpful analogy for many of the phenomena that involve the interference of light waves.

During the nineteenth century wave theory advanced greatly. Augustin Fresnel refined Huygens' idea that an advancing wavefront can be thought of as made up of a series of emitters of new wavelets, by incorporating the principle of interference (Fig. 2.1b). This was particularly helpful in explaining diffraction (the behaviour of light near edges and apertures) which is important in understanding the limitations of lenses. The question of what constituted the waves that make up light was addressed by James Clerk Maxwell, who showed that they could be described as transversely oscillating electrical and magnetic fields that propagated at a finite speed (Fig. 2.1d). Later, in 1888, Heinrich Hertz confirmed Maxwell's idea of the existence of electromagnetic radiation by producing and measuring it. We now accept that light occupies a small waveband (wavelengths between 0.4 and 0.8×10^{-6} m) in an electromagnetic spectrum that extends from γ-rays (10^{-13} m) up to radio waves with wavelengths of many kilometres.

There were still phenomena that wave theory could not explain. One in particular, the photoelectric effect in which light causes electrons to be emitted from metal surfaces, seemed to require a theory in which light interacted with matter as discrete packets of energy. This led Albert Einstein to propose, in 1905, a quantum theory of light which incorporated elements from both wave and corpuscle ideas. Light, according to this scheme, consisted of massless particles whose energy was related to their vibration frequency according to the expression $E = h\nu$, where h is Planck's constant (which has the magnificent value of 6.63×10^{-34} Joule-seconds; Max Planck had introduced the beginnings of quantum theory to explain black-body radiation in 1900), and ν is the frequency of the radiation (for green light, about 6×10^{14} Hz). The minuteness of this quantity of energy, 4×10^{-19} Joules, can be illustrated in mechanical terms; it is the amount of energy liberated by dropping a mass of 4 ng (10^{-9} g) from a height of 1 μm (10^{-6} m). The detailed behaviour of photons remains deeply mysterious, even to physicists. When they interact with matter, as, for example, when they are absorbed by rhodopsin molecules, they behave as discrete packets of energy that cannot be sub-divided, but when travelling through space they can behave as though they are divisible. In a famous repetition of the Young's slit experiment, in which light levels were so low that no more than one photon could possibly have passed through the slits at any one time, a diffraction pattern was formed beyond the slits that was the same as that formed at high light levels. The implication has to be that single photons passed through *both* slits, and interfered with themselves. This is not, on the face of it, consistent with indivisibility of energy, and indeed that idea in its simplest form has been abandoned. Modern ideas are couched in terms of the probabilities of capturing a photon in a particular location, rather than its actual energy distribution. Of the various gnomic utterances on this subject, one of the best comes from W.L. Bragg, of X-ray diffraction fame: 'Everything in the future is a wave, everything in the past is a particle'. The reader who needs to know more should consult a recent optics textbook such as Hecht and Zajac (1997). For the purposes of this book, however, we are mainly concerned with the interactions of photons with matter, when they do behave as countable, indivisible packets of energy, and we will not worry too much about the intimate details of their behaviour in transit.

The ray, wave, and photon descriptions of light are not alternatives, and at the end of the day the photon description has to subsume the other two, just as the Huygens–Fresnel wave theory encompassed the earlier corpuscle-ray theory. However, in the same way that Newtonian mechanics provides a much simpler and more compact way of dealing with ordinary macroscopic events than the more complete theory of relativistic quantum mechanics, so it is often more convenient to deal with light by the simpler partial descriptions. So for our purposes image formation by lenses and mirrors is adequately analysed by geometrical (ray) optics; wave optics are needed to deal with the diffraction limit to the performance of lenses, the behaviour of narrow waveguides such as photoreceptors, the behaviour of multilayer mirrors and diffraction gratings, and wavelength and polarization properties of light; and the photon description is needed to explain the way photoreceptor performance degrades at low light levels when the number of photon 'hits' is inadequate to provide a good statistical sample of the image.

Light intensity

The amount of light available for vision has important consequences for what we are able to see: all aspects of vision degrade as the light gets poor, for reasons explained in Chapter 3. It also affects the evolution of eyes for different light environments: nocturnal and deep-sea animals tend to have particularly large eyes so that they can capture as many photons as possible from the surroundings. On a bright day, the number of photons reaching the earth's surface within the visible range is about 10^{20} per second per square metre. This seems a very large number, given that photoreceptors are capable of detecting single photons, but when one remembers that the dimensions of a photoreceptor are measured in micrometres, and its cross-sectional area in square micrometres rather than square metres, a factor of 10^{12} disappears straight away, and the numbers become more manageable. Bright moonlight is about a millionth as bright as sunlight, and overcast starlight is about ten thousand times dimmer still (Table 2.1). These extremes represent the total range over which human vision is useable – an

Table 2.1

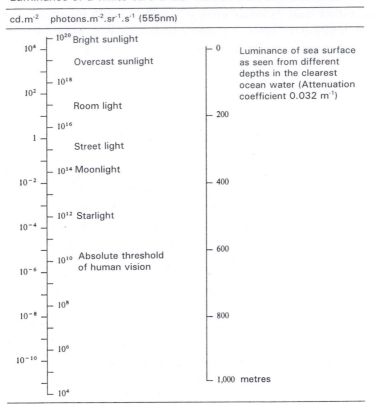

Luminance of a white card under various illumination conditions

| cd.m^{-2} | photons.m^{-2}.sr^{-1}.s^{-1} (555nm) | |

overall span of 10^{10}. At the lower limit, when we can just about see to move if thoroughly dark-adapted, the rate of photon capture is very low indeed: about one per receptor per hour. Individual photoreceptors are capable of giving a satisfactory signal over an intensity range of about 10^5, so supplementary gain control mechanisms, including iris mechanisms and pooling between receptors, are needed to extend the working range in both directions.

In any one scene, the intensity range is nothing like as great as this. Even black velvet reflects about 2 per cent of the incident light, so the maximum brightness range the eye will encounter is a factor of 50. One of the main jobs of the dark and light adaptation mechanisms of the retina is to ensure that under any particular lighting conditions the working range of the retina is limited to this 50-fold intensity range, so that the full processing capacity of the retina is used to register the scene. As the illumination level changes (at dawn or nightfall, for example) the entire range has to shift to a new central intensity level. In this way we are able to see a fully detailed scene in bright daylight, or in roomlight a thousand times dimmer, with very little difference in the perceived result.

Even in the clearest ocean water, blue light (which is absorbed least) is reduced by a factor of 10 for every 70 metres depth, meaning that the human threshold is reached at a depth of 700 metres. Fish with much larger pupils, and some crustaceans with superposition optics (Chapter 8) may be able to see down to 800 or 900 metres, but below that there is effectively no light from the sun. Many animals at this depth do have eyes, but the source of light they use is either their own luminescence or that of other animals. There is a surprising amount of bioluminescence at a depth of 1000 metres, where animals glow or flash to communicate, to seek food, or as a surprise defence. In murky coastal waters light is attenuated much more rapidly, so that little is available after a few tens of metres. There is little bioluminescence either, and the turbidity reduces its value in communication.

Box 2.1 Measuring intensity

Intensity itself is a rather vague term, and it is important to be clear whether we are referring to a source that emits light (where the appropriate terms are luminance or radiance) or a receiving surface (units are illuminance or irradiance). The reason that there are two sets of terms in each case is that they measure light in quite different ways. The first (photometric) system is based on humans as detectors and has its roots in comparisons made in the nineteenth century between different light sources and a 'standard candle'. This may seem archaic but it is still in use; however, the standard is now no longer a candle, but 1 square centimetre of a black-body radiator at the freezing point of platinum. This, of course, is not easy to set up, so most calibrations come from 'secondary' standards, usually carefully calibrated tungsten light bulbs. The second (radiometric) system is based on physical energy measurements (watts, photons per second) that can be traced to universal constants. One important advantage of the radiometric system is that it can take differences in wavelength into account; the luminance system compares all sources of light to a subjective 'white', which may be adequate for some human studies, but is of much less value when studying other animals with vision that is spectrally quite different from ours.

Figure 2.2 illustrates a surface emitting light (left) and one receiving light. Appropriate photometric and radiometric definitions are given in the table below. Let us consider first a radiometric system based on photon numbers. To specify the *radiance* of a surface we need the numbers of photons emitted per unit area per second. Since these are being emitted into the whole hemi-

Box 2.1 Measuring intensity (*contd.*)

sphere in front of the surface, we also need to specify the size of the cone over which the photons are being measured. The appropriate unit here is the steradian or unit solid angle, which is defined as a conical sector of a sphere in which the area of the spherical surface is equal to the square of the radius. Since the area of a sphere is $4\pi r^2$, it follows that 4π steradians make up a complete sphere, or put another way, 4π steradians surround a point. The angular width of a steradian is 65.5° (not the same as a radian, the two-dimensional equivalent, which is 57.3°). Thus the full units of radiance are photons per second per square metre per steradian, or $photons.s^{-1}.m^{-2}.sr^{-1}$. If we are concerned with monochromatic light, those units are sufficient, but if the light is spectrally complex it is also necessary to specify how much of the spectrum is involved. This can be done by breaking up the spectrum into units of wavelength (typically nanometres) and adding nm^{-1} to the preceding definition. The total photon radiance is then given by the integral

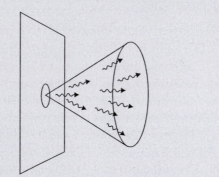

Radiance	*Irradiance*
$photons.s^{-1}.m^{-2}.sr^{-1}$	$photons.s^{-1}.m^{-2}$
or	
$photons.s^{-1}.m^{-2}.sr^{-1}.nm^{-1}$	$photons.s^{-1}.m^{-2}.nm^{-1}$
or	
$W.m^{-2}.sr^{-1}$	$W.m^{-2}$
or	
$W.m^{-2}.sr^{-1}.nm^{-1}$	$W.m^{-2}.nm^{-1}$

Luminance	*Illuminance*
$cd.m^{-2} = lm.m^{-2}.sr^{-1}$	$lm.m^{-2}$

Fig. 2.2 Radiometric and photometric units applicable to light emission and light reception. For details see text.

Box 2.1 Measuring intensity (*contd.*)

across the spectrum of all the spectral elements. For the receiving surface the *irradiance* is the radiant flux (photons per second) per unit area, so its units are photons.s^{-1}.m^{-2}.

A radiometric system using energy units (watts = joules per second) is essentially the same as the photon number system, except that the units are watts (or microwatts) rather than photons per second. The conversion factor is Einstein's equation, given earlier: $E = h\nu = hc/\lambda$, where h is Planck's constant, ν is frequency, c is the speed of light (3.10^8 m.s^{-1}) and λ is wavelength (in metres). For photons in the yellow-green region of the spectrum (555 nm) this works out as 3.6×10^{-19} joules. Thus one watt of yellow-green light is equivalent to about 2.8×10^{18} photons per second.

The luminance system is slightly different because it relies on the candela (cd) as a unit of luminous intensity (a standardized equivalent of the old 'candle power') which incorporates a solid angle in its definition. A point source of one candela emits 4π lumens (lm) of luminous flux, i.e. one lumen into each steradian surrounding the point. Thus an extended source (such as a TV screen) which has a *luminance* of L candelas per square metre, produces a flux of L lumens per steradian per square metre of emitting surface. The *illuminance* of a receiving surface has the units of lumens per square metre, which are also known as lux. (There is a wonderful collection of archaic terms for intensity: stilbs, apostilbs, phots, nits, foot-lamberts etc. Here we stick to SI units as far as possible.) The lumen like the watt is a measure of power, and the two are interconvertible. For light of the most visible wavelength in daylight (555 nm) 1 watt is equivalent to 682 lumens. At the same wavelength one lumen is equivalent to 4.09×10^{15} photons.

If we want to know how much light a surface receives from an emitting surface at a distance d, we can do this by expanding the definition of solid angle in the luminance units. Suppose the emitting surface produces L lm.m^{-2}.sr^{-1}. The solid angle involved here is the area of the receiving surface, divided by d^2, the square of the radius of the sphere of which the solid angle is a part (see above). Thus the definition of solid angle contains within it the better known inverse square law. If the area of the emitter is A_e and the receiver A_r, then the flux (F) at the receiving surface will be:

$F = LA_eA_r/d^2$ lumen,

and the illuminance (I) will be:

$I = LA_e/d^2$ lux.

Radiance and irradiance are similarly related.

Contrast

In general, we and other animals are not particularly interested in the absolute luminance of objects, but in the differences in luminance that define their parts. We need to be able to recognize objects for what they are under a wide range of lighting conditions, so the absolute light level actually needs to be removed, as the visual information is processed. The feature of objects that we need to register is their contrast, which is a measure of the extent to which one part differs from another. For two surfaces whose absolute luminances are L_1 and L_2, the contrast (C) is given by:

$$C = (L_1 - L_2)/(L_1 + L_2).$$

The beauty of contrast, defined in this way, is that it is a property of the object we are looking at, not the lighting conditions. Suppose L_1 and L_2 are two surfaces that reflect different proportions of the light that reaches them, so that the luminance of L_1 is 2 units and L_2 is 1 unit. The contrast, from the equation, is 1/3. If the light shining on them increases a hundred-fold, the contrast will be 100/300, which is still 1/3. Contrast varies between 1, if one surface is completely dark, to 0 if the surfaces have the same luminance.

A number of processes in the retinae of animals ensure that we see contrast rather than raw luminance. Adaptation mechanisms of various kinds mean that the signal passed to the brain is more or less independent of overall light level. The centre-surround organization of ganglion cells means that they signal differences in brightness between adjacent parts of the image, rather than 'spot' values of intensity. There must still be a few neurons that measure intensity to tell us whether it is night or day, but that is not the main job of vision.

Wavelength and colour

The range of wavelengths (λ) visible to humans lies between 400–700 nm (0.4–0.7 μm), with some sensitivity up to 800 nm. This range encompasses the colours of the spectrum famously described by Newton as violet, indigo, blue, green, yellow, orange, red in increasing order of wavelength (Fig. 2.3a). Most people are unhappy with indigo as a distinct colour, but we can all agree about the rest. For many other animals, including birds, fish and insects, the spectrum extends into the ultraviolet range from 400 to about 320 nm. Many flowers have striking markings in the ultraviolet that we cannot see, and which are for the benefit of pollinating insects (Figs 2.3d and 2.4). Some fish and butterflies have visual pigments with maximum sensitivities up to 60 nm further into the red than human visual pigments, so they can see into what, for us, would be the near infra-red. Beyond this, in the micrometre range of wavelengths, is the infra-red radiation given off by hot bodies. Some snakes can make use of these wavelengths for a form of thermal imaging. This involves temperature sensitive nerve endings

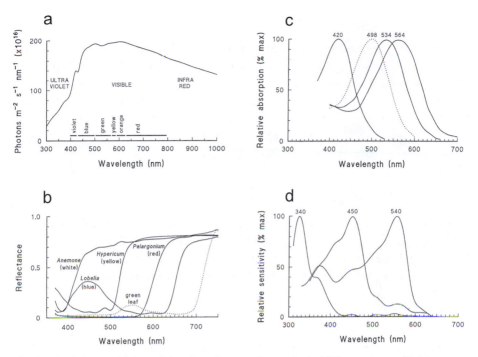

Fig. 2.3 Environmental light and the photopigments that receive it. (a) The spectrum of light reaching the earth's atmosphere from the sun. Note that the visible spectrum occupies the region where photons are most abundant. Data from Lythgoe (1979). (b) The spectral reflectances of four flowers and a leaf. The flowers are illustrated in Plate 1. Note that the anthocyanin colours of the red, yellow, and white flowers all act as long-wave passing cut-off filters. The same is true for the blue, but it is the secondary peak at 450 nm that we see; the long-wave reflectance is too far into the red. The leaf reflects a little in the green (it is the job of leaves to absorb not reflect) and powerfully in the infra-red which is not visible to our eyes. Curves courtesy of Daniel Osorio. (c) The absorption spectra of human rods (dotted) and the three cone types. It is possible to get a rough idea of how much a particular colour would stimulate each cone type by seeing how much overlap there is between the reflectance curve [e.g. (b)] and the absorption curves. Data from Lythgoe (1979). (d) Spectral sensitivity curves of bee photoreceptors. These are essentially similar to the human curves except that they were measured electrophysiologically, rather than by absorption. Note that they extend into the ultraviolet, and are more evenly spaced than the human cone curves. Data from Menzel (1979).

in special pits near the eyes, not the eyes themselves, and visual pigments are not involved. Snakes, which are cold-blooded, use this sense to detect and home in on warm-blooded prey such as rats and mice. The only other animals known to have special detectors of infra-red radiation are certain beetles (*Melanophila*), which approach forest fires from distances of many kilometres. Their larvae are dependent on wood killed by fire (Schmitz and Bleckmann 1998).

Objects in the world around us reflect different wavelengths of light to different extents, and so the wavelength distribution in the light from these objects can provide a valuable clue to their identity (Fig. 2.3b). Leaves reflect most light in the range 500 to 600 nm, blue flowers between 350 to 500 nm, ripe fruit 550 to

Fig 2.4 Ultraviolet markings on flowers and butterflies. *Left*: marsh marigolds (*Caltha palustris*) seen by man as uniform yellow(above), have dark centres in the UV. *Centre*: the yellow butterfly *Phoebus rurina* (male) has brilliant UV markings at the base of the forewings. *Right*: *Bidens* and *Coreopsis* flowers in white and UV light. Redrawn from photographs in Eisner *et al.* (1969).

600 nm, blood 600–650 nm. Being able to analyse in some way the spectrum of light reaching the eye provides a useful tool for classifying different objects.

It is important to recognize that colour and wavelength are not the same. Wavelengths themselves are colourless, and the colours we see are the subjectively perceived result of our wavelength analysis. In the language of philosophy, subjective colours (red, green etc.) are *qualia*, whose nature we cannot demonstrate to others. We may all agree that blood is red and leaves are green, but that does not guarantee that we all see the same colours with our mind's eye (ask yourself what colour a red-green colour-blind person sees when you see orange). It merely says that we agree on their wavelength distributions. It was a surprise in the 1960s, when it first became possible to measure the sensitivity of single cones in the eye to different wavelenths, to find that none of them was sensitive specifically to 'red' wavelengths (longer than 600 nm). The cone closest to 'red' is most sensitive to a wavelength of about 564 nm, which corresponds to a spectral colour of yellowish-green (Fig. 2.3c). Red, of all colours, whose vividness is so impressive, has no special receptor! Increasing redness is represented in the cone signals as a decrease in the output of the 564 nm cones, and an even greater decrease in the output of the 534 nm cones, so that for 'true' red (wavelengths greater than 650 nm) only the 564 nm cones are active.

Confusion arises because we do use our colour names to describe spectral wavelengths. Thus a wavelength of 580 nm is yellow. However, an identical yellow is produced by an appropriate mixture of light of 620 nm (red) and 540 nm (green) wavelengths. The colour we see depends on the relative stimulation of our three cone types, and in this case the pure wavelength and the mixture give the same stimulation ratios. Perceived colour is thus not an accurate

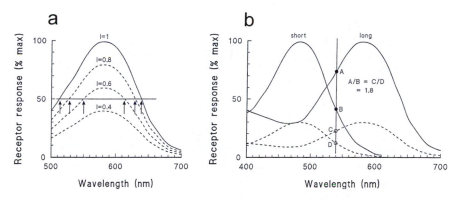

Fig 2.5 At least two visual pigments are needed for colour vision. (a) With only one pigment the response of a receptor does not distinguish between intensity and wavelength. A 50 per cent response could have been produced by any of the arrowed combinations. (b) With two different visual pigments the ratio of stimulation (A/B or C/D) is specific to a particular wavelength, and unaffected by intensity level.

guide to spectral composition. There are also many colours we see that do not have corresponding single spectral wavelengths: purple for example is a mixture of long (red) and short (blue) wavelengths. Colour science is an important but complex subject, and as we are more concerned here with animals whose colour vision system is not like our own, the interested reader should consult a text such as Mollon and Sharpe (1983).

If we are uncertain about the relationship between perceived colour and wavelength discrimination mechanisms in our own species, we should obviously be even more cautious in thinking about what sort of colour vision other animals have. We can be certain, however, that a great many animals do have it. The ability to discriminate lights with different wavelength distributions depends on an animal possessing at least two visual pigments with different spectral sensitivities. Then, as Fig. 2.5 shows, spectral colours of different wavelengths will give unique *ratios* of stimulation of the two pigments, independent of the total stimulation; that is, the overall level of illumination. With only one visual pigment wavelength and intensity cannot be disentangled from each other, and colour vision of any sort is impossible (There is an alternative, which is to have one visual pigment and several colour filters. There are indeed such filters in some photoreceptors (the coloured oil droplets in the retinae of birds and reptiles, for example) but their function seems to be to 'sharpen up' the spectral sensitivities of the cone pigments, rather than to create a colour vision system from a single photopigment.) Thus if an animal possesses two or more visual pigments in its eye, there is a *prima facie* case for thinking that it has colour vision of some kind. The great majority of arthropods and vertebrates do indeed have at least two and typically three visual pigments. Some have even more, the record being 15 in stomatopod crustaceans (Marshall and Oberwinkler 1999).

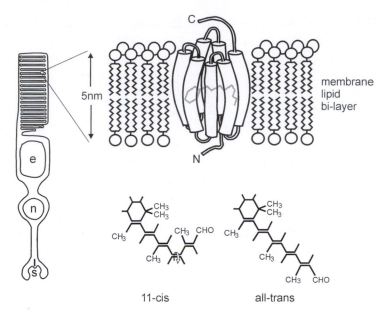

5nm

membrane
lipid
bi-layer

e

n

s

CH₃
CH₃
CH₃
CH₃ CHO

CH₃

CH₃

11-cis

CH₃
CH₃
CH₃

CH₃

CH₃ CHO

all-trans

Fig 2.6 *Left*: digrammatic section of a vertebrate rod, showing the discs of membrane that contain the photoreceptor molecules. s, synapse; n, nucleus; e, ellipsoid (mitochondria). *Upper right*: diagram of a rhodopsin molecule in the membrane, showing the seven helices that enclose the chromophore group, retinal. C and N are the carbon and nitrogen termini of the opsin protein. *Lower right*: the retinal molecule in its unstimulated (11-cis) and stimulated (all trans) form. The light-sensitive double bond lies in the plane of the membrane. After Lythgoe (1979).

There are two parts to a visual pigment molecule. The part that receives the photon is the *chromophore*, one of four close relatives of vitamin A. These have a long chain of alternating single and double bonds, in which the bond between the 11th and 12th carbon atoms reacts to the capture of a photon by changing from the *cis* to the *trans* configuration (Fig. 2.6). This then initiates a series of biochemical events which results in the closure of sodium channels and a hyperpolarization of the cell (in vertebrates) or an opening of sodium or calcium channels (in most invertebrates) and a consequent depolarization. The wavelength range that a photopigment molecule responds to best depends partly on which of the four chromophores is present, and partly on the structure of the protein molecule (the opsin) that surrounds the chromophore (Fig. 2.6). It is now known that a handful of amino acids in the region around the chromophore can 'tune' it, so that it responds best to photons of higher or lower energy. Thus, colour vision systems contain photopigments that possess either different chromophores, or different opsins, or both. A good account of the photochemistry of vision can be found in Rodieck (1998).

'True' colour vision is usually taken to mean that an animal can use or learn to use not just the wavelengths that correspond to the peak sensitivities (λ_{max}) of the visual pigments, but also intermediate wavelengths and wavelength combinations, by making use of stimulation ratios. Our colour vision is like this, and so is the colour vision of bees (Fig. 2.3d) which can be trained to a wide variety of coloured stimuli. There are, however, simpler systems, referred to as 'wavelength specific behaviours' where the outputs from the different photoreceptors seem to drive behaviour directly. A good example of this is the cabbage-white butterfly *Pieris brassicae* where the 'open space' escape reaction is driven by wavelengths around 370 nm in the ultraviolet, the feeding reaction to wavelengths around 460 nm and also 600 nm (i.e. flower colours in the blue and red), and egg-laying by green wavelengths around 540 nm (Scherer and Kolb 1987). These wavelengths correspond closely with the peak sensitivities of butterfly visual pigments. However there are also indications that wavelength mixtures can be effective, and it seems likely that butterflies have some 'true' colour vision as well as these wavelength specific behaviours.

Polarization

Polarization is a property of light that we are unable to detect, but whose use is commonplace in the animal kingdom. As indicated in Fig. 2.1d, the electric field of a photon lies in a particular plane, and a photon will only excite a photopigment molecule if the direction of this vibration, and the orientation of the excitable double bond in the photopigment molecule (the 11-cis bond of the chromophore group) lie in the same plane. In the discs that make up the photoreceptors of vertebrate rods and cones the photopigment molecules lie in a plane perpendicular to the incoming light, but in all possible orientations within that plane (Fig. 2.7b). That means that the receptor cell has no means of knowing what the direction of the electric vector of the photon it received might have been. In the microvillous receptors of invertebrates the situation is different. A long tube, such as a microvillus, covered with a photopigment-bearing membrane has, just from its geometry, a 2:1 preponderance of chromophore groups aligned parallel to its long axis (Fig. 2.7c and d). This means that microvillous (or 'rhabdomeric') receptors have a built-in capability to respond selectively to light polarized in a particular plane. To make a system that can actually determine the direction of polarization of the light reaching the eye requires two or more groups of receptors with their microvilli aligned in different directions, and a neural system that is able to work out the ratio of the responses. This is a very similar problem to that of colour vision (Fig. 2.5) and no doubt the neural solution is similar. Many insects and some crustaceans are capable of this kind of analysis.

Light from the sun is unpolarized; that is to say, it contains photons whose electric fields are in all possible orientations. However, two processes – scattering and reflection – distinguish between photons with different electric field direc-

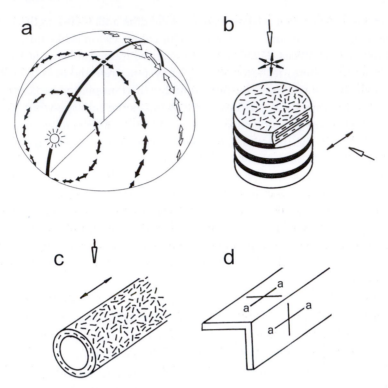

Fig. 2.7 Polarized light and its reception. (a) The pattern of polarization in skylight. The E-vector directions are concentrically arranged around the line joining the sun to the 'anti-sun' 180° away. The polarization is most intense at 90° from the sun (*open arrows*). When the sun is not visible an insect can infer its direction from small regions of the polarization pattern. (From Rossel 1989). (b) The random distribution of chromophore molecules in a rod disc means that a rod cannot distinguish between photons with their E-vectors in different planes, when light reaches the disc from its normal direction. Light from the side, however, is only absorbed if polarized parallel to the disc membrane, because that is the orientation of the chromophore molecules (see Fig. 2.6). *Open arrows*, light direction; *filled arrows*, E-vector direction. (c) The finger-like microvilli of invertebrate rhabdoms have a preponderance of chromophore molecules aligned parallel to their long axes. This is most easily demonstrated with the square section in (d). Here it is clear that there are twice as many molecules aligned in the direction a–a than in either of the other orthogonal directions. In some microvillous receptors specifically involved in polarized light reception the molecules are not randomly arranged in the membrane, but specifically aligned with the microvillar axis.

tions, and result in polarized light. Both processes are useful to animals. As the sun's rays pass through the atmosphere fine particles scatter out blue light, and also preferentially scatter light polarized in a plane at right angles to the ray-path from the sun (Fig. 2.7a). This results in a pattern of polarization in the sky which is determined by the sun's position, and even if the sun is obscured by cloud the polarization pattern largely persists. This pattern can thus be used instead of the sun as as a navigation aid, a role which has been thoroughly demonstrated in bees and ants (Rossel, 1989) and suspected in many other animals.

Non-metallic reflecting surfaces, water for example, also polarize light. At one particular angle (Brewster's angle; 53° for water) the polarization is complete, so that the reflected light is all polarized in one direction (parallel to the surface), and the transmitted light is all in the plane at right angles. The glare from water can be a nuisance to us, so we often cut it out with polaroid sun-glasses that selectively absorb light polarized parallel to the water surface (Fig. 2.8). Some water bugs, however, make use of this polarized reflexion specifically for the purpose of finding water when their particular pool dries up. Rudolf Schwind (1983) used a sheet of polaroid to mimic a water surface, and found that water boatmen (*Notonecta*) would crash land onto the polaroid with the same enthusiasm as they would dive into a real water surface. Both bees and water bugs have special regions of the eye containing receptors with microvilli aligned in particular directions, in a pattern apparently designed to extract the necessary polarization information.

Polarization vision has also been implicated in communication. Both cuttlefish (Mollusca) and mantis shrimps (Crustacea) have specific patterns on conspicuous parts of the body that are only visible to a polarization-sensitive viewing system. Both animals are known to have polarization vision, and indeed mantis shrimps have been shown to be able to learn polarization patterns (Marshall *et al.*, 1999).

Fig. 2.8 Examples of natural polarization. *Left*: photographs of a water surface and a matt grey card with a polaroid filter aligned parallel (above) and perpendicular (below) to the water surface. *Right*: polarization as pseudo-colour. Three leaves (a matt sage leaf, a bay leaf, and a shiny cotoneaster leaf) photographed through polaroid filters as in the left-hand photographs. Note that the brightness order of the leaves reverses as the polaroid cut out the reflection from the shiny leaves.

Summary

1 Light can behave as rays, as waves or as streams of particles. For most optical purposes a description in terms of rays is adequate, but several phenomena including the resolution of images can only be explained by wave interference. At low light levels the quality of vision depends on the statistics of particle (photon) numbers.

2 Human vision extends over an intensity range of about 10^{10}. In general visual systems detect contrast rather than intensity, where contrast is the difference in intensity of two surfaces divided by their sum. It depends on the reflectances of objects rather than the intensity of illumination.

3 Objects reflect light of different wavelenghts to different extents, and this is the basis of colour. Colour vision requires at least two visual pigments that are maximally sensitive to different wavelengths.

4 Polarization vision is common in animals. Light is polarized by scattering in the atmosphere, and the pattern produced can can be used as a navigation aid. Water surfaces and other non-metallic reflectors also polarize light. Detection requires that the photopigment molecules are appropriately aligned in the photoreceptor membrane.

3 | What makes a good eye?

Fundamentals

Eyes are unique amongst the sense organs because we know enough about the physics and chemistry of vision to be able to say with some certainty why they are built the way they are. Of course, they were not designed, as one would design a camera or telescope, but evolved over millions of years. Nevertheless, both evolution and technology have to obey the same set of physical rules. Image-forming lenses, for example, have to be made using the principle of refraction by a transparent high refractive index material, whether the lens evolved in an octopus or fish, or was designed by Leitz or Nikon. The differences come in the materials: biological lenses are generally constructed from protein rather than glass, and mirrors are made from guanine multilayers rather than silver. It is chemistry rather than physics that distinguishes biology from technology. In this chapter we explore these physical constraints on eye evolution. We will make the fairly bold claim that it is sensible to approach eyes in essentially the same way that an optical engineer might evaluate a new video camera. We can say what most of the components are for and how well they are likely to perform, and also establish criteria for judging the performance of an eye as a whole. Thus this chapter is intended as something of a tool kit for interpreting eye structure, and for providing a basis for comparing the performances of the different types of eye that will be the subject of later chapters.

Eyes supply information about the nature of the light distribution in the environment. For a hawk this information needs to be very fine-grained, but for a flat-worm it can be coarse. Although we cannot say that the flatworm's simple pigment cup eye is less successful, in an evolutionary sense, than the hawk's, we can nevertheless say that the hawk's eye is better, because of the much greater quantity of information it is capable of supplying to its bearer. If we are to employ 'information supply', albeit loosely, as our basis for judging the quality of an eye, what yardsticks should we use? We will leave aside for the moment the capacity to distinguish wavelength and plane of polarization; these are features more of the molecular organization of the receptors than of the structure of the eye itself (see Chapter 2). It is generally agreed that there are two features of an eye's function that between them summarize its performance, and which are independent of the eye's optical type. These are *resolution* and *sensitivity*. By resolution we mean the precision with which an eye splits up light according to its direction of origin. This is a combination of the quality of the image provided by the optics, and the fineness of the mosaic of retinal detectors. Sensitivity refers to the ability of an eye to get enough light to the receptors for them to make full use of the eye's potential resolution. For animals living in dim environments, sensitivity is every bit as important as resolution.

Before examining in detail the features of an eye that make for good performance, we will first look briefly at the way that resolution and sensitivity interact, and the reasons why both are important. Figure 3.1 is an imaginary eye with rather poor resolution and a dim image, intended to show in exaggerated form the problems that all eyes face. Two point sources of light outside the eye are brought to a focus on the retina by the lens, where they give rise to distributions of light that are no longer point-like, but blurred and spread out over several receptors (blur circles). There are many possible reasons for this spread. For example, the optical system might fail to bring all rays to a single focus (aberration), or light might be scattered by the media of the eye. Even if the eye is perfect in these respects, there remains a fundamental source of blurring known as *diffraction*, which is inescapable, and arises from the wave nature of light (see Figs 2.1 and 3.5 below). This will be discussed later in the chapter, but basically the smaller the aperture of the eye compared with the wavelength of light, the worse the problem is, so that in the tiny optical systems of insect compound eyes, for example, diffraction is particularly serious (Chapter 7). The degree of blur resulting from these defects limits the quality of images of all kinds, and in doing so also establishes how fine the retinal mosaic should be. There is no point in the retina having receptors much smaller than the blur circles that make up the image. Roughly speaking, all the information contained in the image is extracted when two receptors occupy the half-width of the light distribution in a point source image, more or less as shown in Fig. 3.1. Thus the poorer the image quality the fewer the number of receptors needed to take in all the information the image offers. A coarser mosaic than this will waste image detail (and can be said to 'undersample' the image), and a finer mosaic will have more receptors than necessary ('oversampling'), so there is a clear optimum. Most eyes do indeed show this expected match between image quality and retinal 'grain'.

Figure 3.1 also illustrates the effect of low light levels, by showing (black dots) how many photons each receptor captures from the image. Light is quantal, and the smallest packet of light energy, the photon, is indivisible when it is caught by a rhodopsin molecule: it is either present or not present (Chapter 2). This means that at low light levels there is much statistical uncertainty, represented in Fig. 3.1 as the variable numbers of photon captures in receptors supposedly each receiv-

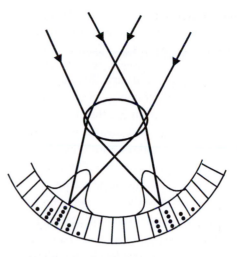

Fig. 3.1 Limits to resolution. An imaginary eye showing the blurring of the image by imperfect optics, the way the image is sampled by the retinal mosaic, and the uncertainty resulting from low numbers of photons (dots).

ing the same average amount of light. If more photons were available then the photon number distributions would come increasingly to resemble the optical distributions, but if fewer were present the situation would become much worse, with only an occasional photon reaching any of the receptors that image[1] each point source. Thus low light levels corrupt the image by introducing uncertainties in photon numbers. One can think of this producing a kind of statistical blur, which adds to the blur caused by diffraction or imperfect optics, and which has similar deleterious effects on the eye's ability to resolve. At low levels this often means that it is better to employ large receptors, in order to get a reasonable statistical photon sample, than it is to have small receptors to sample the image finely. This trade-off between resolution and sensitivity is one we shall meet repeatedly, particularly in animals that have to operate over a range of light levels.

We have seen that both wave and particle aspects of light affect eye performance. Its wave nature imposes a fundamental limit to image quality, and, as we will see later, to receptor size as well; and its quantum nature determines the certainty with which light can be measured. With these constraints in mind, the rest of this chapter will be used to explore the features of eyes that enable and limit their capabilities.

---

[1] Throughout this book we will use 'image ' both as a noun and as a verb meaning 'to form an image'. 'Focus', as a verb, will be used to mean altering the position of the image to bring objects at different distances to a focus, i.e. to effect accommodation.

Resolution

The retinal sampling frequency

The two features of an eye that set a limit to the detail that can be resolved in bright light are the fineness of the receptor mosaic and the quality of the image. How can we best compare the effects of these rather different attributes? It turns out that one of the best measures that can be applied to both is their capacity to resolve a grating of dark and light bars. In the case of the receptor mosaic, it is a well-established finding that a grating can be properly resolved if the image of each adjacent dark and light stripe falls upon a separate receptor. This means that the period (the distance between the centres of two adjacent dark or light stripes) of the finest resolvable grating in the image is equal to twice the receptor spacing.

When dealing with objects outside the eye and images within it, it is often most convenient to deal with angles rather than distances, as the same angular measurements apply to both. In single-chambered eyes like our own there is always a point in the eye called the *nodal point* that rays pass through without being bent by the lens. For example, in an eye that forms an image with a simple curved cornea the nodal point will be at the centre of curvature of the corneal

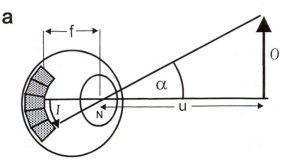

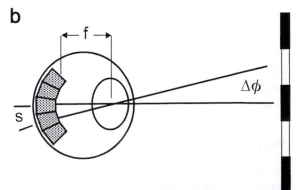

Fig. 3.2 Objects and images. (a) A distant object subtends the same angle α inside and outside the eye when the ray passes through the nodal point N. The focal length (f) is the distance from the nodal point to the image. (b) The finest grating that an eye can resolve has an angular period of $2\Delta\phi$, where $\Delta\phi$ is the inter-receptor angle (s/f) at the nodal point, and s is the separation of the receptor centres.

surface, because rays passing through that point will meet the surface at a right angle, and so will not be bent by refraction. The significance of the nodal point is that one can draw straight lines through it connecting object and image points, and so work out the relative sizes of objects and images directly by the principle of similar triangles [Fig. 3.2(a)]. A small object of size O at a distance U from the eye makes an angle of $\alpha = O/U$ radians at the nodal point (a radian is the angle made by an arc of a circle one radius in length at the circle's centre: 1 radian = $180°/\pi$, or 57.3°), and inside the eye the image I subtends the same angle α at the nodal point. The best definition of the focal length (f), for our purposes, is the distance from the nodal point to the image of a distant point. Then the equation:

$$O/U = \alpha = I/f \tag{3.1}$$

summarizes the relations between object and image, provided the object is a long way away. A particularly important angle, because it determines the fineness with which the image is sampled, is the inter-receptor angle, s/f, where s is the spacing of the receptor centres. The symbol $\Delta\phi$ will be used for this angle [Fig 3.2(b)].

We can now apply eqn (3.1) to the grating resolution of the eye. The finest resolvable grating has a period of $2s$ on the retina. Expressed as an angle in either image space inside the eye or object space outside, this is $2s/f$ radians. It is often more useful to speak of a grating's *spatial frequency* (the reciprocal of the period, in cycles per radian) because the frequency increases as the resolution improves, whereas the period decreases. The spatial frequency with which the retina samples the image is the *sampling frequency*, designated by (greek nu) ν_s. Thus:

$$\text{sampling frequency } (\nu_s) = f/(2s) = 1/(2\Delta\phi) \tag{3.2}$$

This equation suggests that there are two ways to increase the sampling frequency, and so improve the eye's resolution. The focal length f might be increased, or the receptor separation s decreased. It is not possible to decrease s below about 2 μm, because receptors narrower than this become leaky to light, as discussed later. Once this limit is reached, as it is in many animals, the only way to improve matters is to increase the focal length, and this necessarily means having a larger eye.

Table 3.1 shows the resolution of the eyes of a variety of animals, expressed in terms of both the inter-receptor angle $\Delta\phi$, and the sampling frequency ν_s. They range from flatworms with a sampling frequency of about 1 cycle per radian, up to eagles with about 8000 cycles per radian.

The optical cut-off

As the detail in a scene becomes finer, the more difficult it is to resolve; the leaves of distant trees lose their identity in the overall texture. One can think

Table 3.1 The resolution of a selection of animal eyes

Name	Maximum resolvable spatial frequency (cycles per radian)	Equivalent inter-receptor angle (degrees)	Method	Ref.
Aquila (eagle)	8022	0.0036	B,A	1
Man (fovea)	4175	0.007	B,A	2
Octopus	2632	0.011	A	2
Portia (jumping spider)	716	0.04	A	3
Cat	573	0.05	B	4
Goldfish	409	0.07	B	5
Aeschna (dragonfly)	115	0.25	A	2
Hooded rat	57	0.5	B	4
Worker bee	30	0.95	B,A	2
Leptograpsus (crab)	19	1.5	A	6
Pecten (scallop)	18	1.6	B,A	2
Lycosa (wolf spider)	16	1.8	A	5
Littorina (sea snail)	6.5	4.5	A	2
Drosophila (fly)	5.7	5	B,A	2
Limulus (horseshoe crab)	4.8	6	A	6
Nautilus (cephalopod)	3.6	8	B,A	2
Cirolana (deep-sea isopod)	1.9	15	A	6
Planaria (flatworm)	0.8	35	A	2

Methods: A, anatomical; B, behavioural. Where the behavioural methods give a lower resolution than the receptor separation, the behavioural result is used. In vertebrate eyes pooling may result in reduced resolution.
References: 1. Reymond (1985). 2. Land (1981). 3. Land (1985). 4. Charman (1991). 5. Nicol (1989). 6. Land and Nilsson (1990).

of the world, from an optical point of view, as consisting of gratings of a complete range of spatial frequencies. Evidently the highest spatial frequencies, representing the finest detail, do not survive the process of vision. It might be

Fig. 3.3 The contrast transfer function. The graph shows what happens to the contrast of gratings of different spatial frequency when they are imaged by a diffraction limited, but otherwise perfect, lens. As the gratings get finer (higher ν) the contrast in the images decreases until it reaches zero at the cut-off frequency (ν_{co}). The ordinate is the ratio of the image contrast to that of the object. The insert shows that the effect of the lens is to convert a high-contrast object into a lower contrast image.

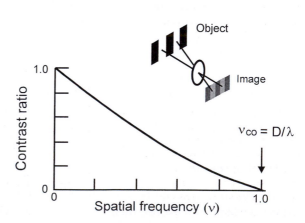

that the retinal mosaic fails to sample them fully, as we have discussed already, but that is not the only reason. The optics of the eye also attenuate, and ultimately cut out the highest frequencies. One of the best ways to illustrate this is by measuring the *contrast (modulation) transfer function* (Fig. 3.3). This is a graph that shows how the contrast of a grating is reduced on passing through a lens, as a function of spatial frequency. [Contrast is defined in Chapter 2: for a grating, it is the difference in intensity of the light bars (I_{max}) and dark bars (I_{min}), divided by their sum, i.e. Contrast = $(I_{max} - I_{min})/(I_{max} + I_{min})$. Dividing by $(I_{max} + I_{min})$ makes contrast independent of the overall light level]. The effect of an optical system is always to reduce the contrast in the image, compared with the object grating that gave rise to it (insert, Fig 3.3), and this reduction is greatest for the highest spatial frequencies. Eventually, as Fig 3.3 shows, a spatial frequency is reached where there is no contrast at all in the image, and this is known as the *optical cut-off frequency* (ν_{co}).

The diffraction limit

It is diffraction that sets the cut-off frequency. Other optical imperfections, not being properly focused for example, may cause contrast to be reduced at all spatial frequencies, but they do not necessarily change the cut-off itself. Diffraction is thus of key importance in understanding both image quality and eye design. It arises from the wave nature of light. When light from a distant point object, such as a star, reaches a lens, the parallel rays are bent by refraction so that they come together at a single focus in the image plane. An alternative, and more accurate, description of the same process is to say that light from the star reaches the lens as a wavefront ('rays' are arbitrary lines at right angles to this front, see

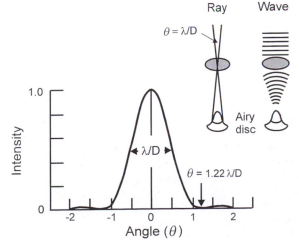

Fig. 3.4 Light distribution in the Airy disc. According to wave optics, the image of a point source is a diffraction pattern known as the Airy disc, which has the intensity distribution shown. Its width depends inversely on the aperture diameter D. The angle on the abscissa is given in multiples of the half-width of the Airy disc, λ/D radians, where λ is the wavelength of light. The inserts show the meaning of θ in terms of ray optics, and the way a point image is formed according to wave optics.

Box 3.1 The origins of the Airy diffraction pattern

A full derivation of the Airy diffraction pattern is available in textbooks of Optics, but it is so important for understanding the limits of vision that an indication of how it comes about is needed. In Fig. 3.5a two rays from the converging wavefront interfere in the region of the focus. These come from two points, X and Y, in the centre of each half of the aperture (the whole aperture can be thought of as made up of a series of similar pairs of points). We then ask the question: 'How far from the focus do we have to go before the image becomes dark?'. This 'first dark ring' can be thought of as defining the edge of the image of the point object. For a point A in the centre of the

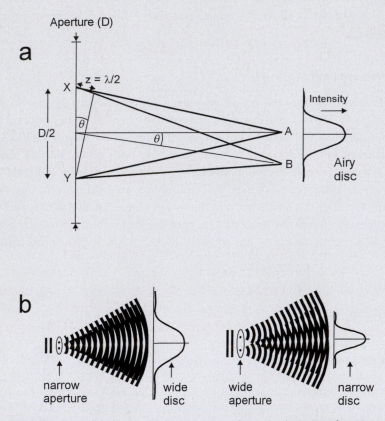

Fig. 3.5 Diffraction and the image. (a) Construction to show why the image of a point source becomes dark away from the axis. When there is a half-wavelength ($\lambda/2$) difference in path length between the rays reaching the image from the two halves of the aperture, destructive interference occurs, and the image at B will be dark. The angle corresponding to the distance AB in the image is θ, which for a circular aperture is equal to 1.22 λ/D (Fig. 3.4). b) Interference patterns illustrating why wide apertures (*right*) produce narrower 'Airy discs' than narrow apertures (*left*).

Box 3.1 The origins of the Airy diffraction pattern (*contd.*)

image the distances from X and Y in the aperture are the same, so the waves will be in phase, they will interfere constructively, and A will be bright. However, at point B the distances are no longer the same, and if the difference in the lengths of the paths from X and Y is equal to half a wavelength of light ($\lambda/2$) the two waves will be exactly out of phase and will interfere destructively, and so B will appear dark. On Fig. 3.5 we can see that the triangle converging on B differs from that converging on A by being tilted through an angle θ and by having an extra short segment z in one of its sides. The length of z will determine what kind of interference (constructive or destructive) occurs at B. z is shared by another triangle containing X and Y and the base of the triangle converging on B, also tilted through an angle θ. The distance between X and Y is half the aperture diameter ($D/2$) so the angle θ in radians is given by $z \div D/2$. If B is to be dark, z must be equal to $\lambda/2$. Substituting this for z gives $\theta = \lambda/D$. This is now the angular position, relative to the centre of the lens, of the dark ring marking the edge of the bright image. It can be converted to distance in the image plane by multiplying by the focal length, as in eqn (3.1). In spite of the simplifying assumptions, this result is very close to that of Airy's complete calculation, which gave $\theta = 1.22 \, \lambda/D$, for a lens with a circular aperture.

Fig 2.1a). On passing through the lens the central region of the wavefront is delayed more than the edge regions, because it passes through more of the optically dense material. The result is that the emerging wavefront is no longer flat, but curved into a part-spherical shape, centred on and progressing towards the focus (insert, Fig. 3.4). At the focus the various parts of the wavefront meet and as they pass through each other they interfere. Components that are in phase with each other will reinforce, whilst those that are exactly out of phase will cancel, giving rise to a pattern at the focus that is not a point (as supposed by ray theory), but a *diffraction pattern*. In the simple case of a point source object and a circular aperture this pattern has a central bright spot known as the *Airy disc* (after its discoverer) and has the form shown in Fig. 3.4.

A convenient measure of the size of the Airy disc is its half-width, i.e. its width (w) at half maximum intensity (Fig. 3.4). This turns out to be almost identical to the distance of the first dark ring discussed in Box 3.1. To a good approximation this is given by:

$$w = \lambda/D \text{ (radians).} \tag{3.3}$$

The larger the value of w, the wider the image of a point, or more colloquially the more blurred the image. Roughly speaking, objects whose images are larger than w will be resolved, but smaller images will not, because they are blurred out. This is reflected in the contrast transfer function (Fig. 3.3) by the fact that the finest resolvable spatial frequency – the cut-off frequency (ν_{co}) – is simply the reciprocal of the Airy disc half-width:

$$\text{cut-off frequency } (\nu_{co}) = 1/w = D/\lambda. \tag{3.4}$$

In other words, the finest grating that the optics can resolve has a period equal to the half-width of the image of a point source.

Equation (3.3) shows that angular image size, θ ($= w/f$), which determines resolution, is *inversely* proportional to aperture diameter. The bigger the lens the smaller the value of θ, and the better the resolution. This is an important and strangely counter-intuitive conclusion. One might think that scaling up an optical system and making the aperture larger would cause the width of the image-disc to grow in proportion; but in fact the opposite is true (Fig. 3.5b). This is why astronomers need big telescopes to resolve small closely spaced stars. By the same token, it is the reason why insect eyes, whose lens diameters are measured in micrometres, resolve so poorly.

By way of example, we can use eqn (3.3) and (3.4) to make a comparison between the theoretical resolution limits of the eye of man and of a bee. The human eye has a pupil about 2 mm wide in daylight, so that for a wavelength of 0.5 μm (blue-green), w comes to 0.00025 radians, 0.014°, or 0.86 minutes of arc. The corresponding cut-off frequency is 70 cycles per degree, which is very close to the sampling frequency of the retinal mosaic, about 60 cycles per degree (ν_{s}, eqn 3.2). The compound eye of a bee, however, has facets that are only 25 μm in diameter. This is smaller than the human pupil by a factor of 80, and consequently the resolution must be 80 times worse, with w about 1.1°. To get a feeling for what this means, your little finger nail covers about 1° with the arm extended. It is easy to use this to imagine how blurred the bee's visual world would be, compared with our own.

Other optical defects

Although diffraction is the ultimate limit to resolution, which can only be improved upon by making the aperture of the eye bigger, there are several other ways that resolution may be compromised. The most important are focus, spherical aberration, and chromatic aberration, and they all occur in animal eyes (Fig. 3.6). Near objects are brought to a focus further from the lens than distant objects, so in large eyes particularly it is important that the optical system can adapt in some way to object distance. This process of accommodation may be accomplished by changing the power of the lens (in man) or by moving the lens

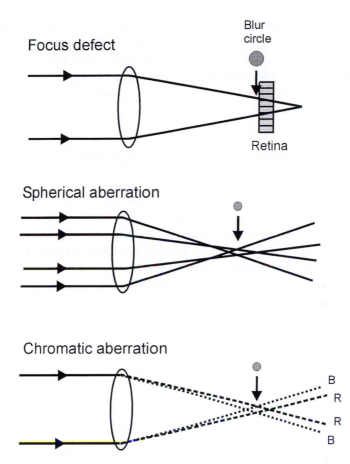

Fig. 3.6 Other optical defects. *Top*: image not in focus on the retina. Middle: spherical aberration, in which outer rays are focused closer to the lens than rays near the axis. *Bottom*: chromatic aberration, where different wavelengths (red and blue) are focused at different distances. In each case the result is a blur circle that adds to the blur due to diffraction.

(in fish). An out of focus image is degraded because point sources produce 'blur circles' on the retina which, like the Airy disc, depress the contrast transfer function (Fig. 3.3).

Spherical aberration is the name given to the blurring that occurs because a simple spherical surface does not bring all rays together at a single focus. Rays furthest from the axis of the lens are refracted too much, and finish up in front of the focus for rays near the axis, again resulting in a blur circle which is larger than the Airy disc. This is potentially a serious problem for biological lenses, but animals get round it in one of two ways. They may make the optical surfaces non-spherical, and indeed the human cornea is not spherical but hyperbolic in shape to avoid just this problem. A common alternative is to make a lens which is not optically homogeneous, as glass is, but which has a gradient of refractive index from the centre (high) to the periphery (low). The result is that the outer zones of the lens refract less than they would in a glass lens, and with the correct gradient of refractive index all imaging rays can be brought to a single point. That fish lenses have this construction, and correspondingly excellent optics, has been known since the studies of Matthiessen in the 1880s (see Chapter 4). The human lens has inherited this design from our fishy ancestors, and corrects its own spherical aberration this way. Thus the human eye has both non-spherical (cornea) and inhomogeneous (lens) correction mechanisms.

Chromatic aberration is caused because short wavelength blue light is refracted more strongly than long wavelength red light. This occurs in biological materials just as in glass, and means that the blue image in the human eye is almost 0.5 mm in front of the red image. No animal eye seems to have emulated the achievement of the early telescope makers in making an achromatic lens by using a combination of materials. However, there are other solutions. Humans evade the problem by using only a relatively narrow range of wavelengths in the middle of the spectrum for high-acuity vision, so that our resolution in the blue end of the spectrum is poor – little more than a 'colour wash'. Some fish are a little more subtle, and have lenses with multiple focal lengths. This ensures that each cone type has an in-focus image for at least a proportion of the light reaching it (see Chapter 4).

Unlike diffraction, where eye size is a virtue, these other problems get worse as eyes get bigger. The reason is that the blur circles caused by aberrations scale with the focal length of the eye, so that, uncorrected, an eye of 1 cm focal length would have blur circles 10 times as large as in an eye of the same design, but with a 1 mm focal length. However, receptors do not in general scale with focal length, but have much the same diameter whatever the size of the eye. Thus the potential resolution of the eye, measured by the inter-receptor angle $\Delta\phi$ (= s/f), should improve as the focal length increases. However, it can only do so if focus defects and other aberrations are minimized, so that blur circles do not get much larger than receptor diameters. For structures with short focal lengths, for example the ommatidia of apposition compound eyes with focal lengths of about 100 μm, these defects are negigible compared with diffraction; no insect needs a mechanism for focusing its eyes. They become noticeable at a focal length of a millimetre, and serious when this reaches a centimetre. In all vertebrates and also the cephalopod molluscs these three kinds of optical problem have been addressed, in one way or another. A contractable pupil is particularly important in dealing with optical defects, as it can be used to strike an appropriate compromise between diffraction (wide pupil) and aberrations (small pupil). This compromise changes with intensity, as bright light favours high acuity, but in dim light the priority is to obtain adequate photon numbers (see also Chapter 5, Fig. 5.10).

Photoreceptor optics

So far, this section has only dealt with resolution in terms of the quality of the optical image, but the ability of the eye to transmit the information contained in the image also depends on the size of the photoreceptors – as well as on their spacing as we have already discussed. If a receptor has a diameter that is narrower than a line in the finest grating that the eye can resolve, then it will be able to measure the intensity (strictly illuminance, see Fig. 2.2) of that line accurately. If, however, if it is much wider, it will swallow up that line and several others, and signal an unresolved average intensity for the grating. For eyes whose function is to resolve well in daylight narrow receptors are therefore essential, and this is indeed what is found. Cones in human eyes are about 2 μm in diameter, which is almost exactly the width of a single line in a just-resolved grating of 70 cycles per degree. In a bee's eye the receptors are also about 2 μm wide, but because the focal length of a facet in a bee's eye is so short (about 100 μm), the angle involved is much larger, just over 1° (α in eqn 3.1). As in humans, this is close to the optical resolution limit imposed by diffraction.

Why should receptors not be narrower still? As we have just seen, an eye's ability to resolve depends, among other things, on the angle subtended by single receptors. As this is equal to s/f radians (Fig. 3.2) it would seem that an eye could be made smaller by reducing its focal length (f), without losing resolution, provided the receptor diameter ($d \approx s$) could be reduced at the same time. This doesn't happen, however. The narrowest receptors in vertebrates and in insects are about 1 μm wide. The main reason for this seems to be that as the width of a photoreceptor begins to get close to the wavelength of visible light (0.3 to 0.8 μm), the receptor is no longer able to hold the light within it by total internal reflection

Fig. 3.7 Receptor optics. (a) In a wide receptor (*left*) light is trapped by total internal reflection. This occurs only up to the critical angle (θ_{crit}), which is given by arcsin(n_1/n_2), where n_1 and n_2 are the refractive indices outside and inside the receptor, typical values of which are 1.34 and 1.36–1.40. (b) In very narrow receptors (diameter <2 μm) the light behaves as a waveguide mode, and has a distribution in which some travels outside the structure. This can be caught by neighbouring receptors (stipple).

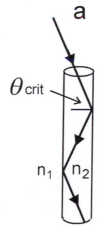

(Fig. 3.7a), and it becomes inefficient and 'leaky'. Like diffraction, this is a phenomenon associated with the wave nature of light. In narrow light-guiding fibres, which is what photoreceptors are, the trapped light forms interference patterns which are known as waveguide modes (Fig 3.7b, Plate 3); these are similar in nature to the standing waves in organ pipes, their acoustic equivalent. The light in these modes is not uniformly distributed, and in particular the single mode found in the narrowest fibres has a substantial part of its energy actually outside the fibre (explanations of this are given by Snyder 1979, and van Hateren 1989). Not only is this light unavailable for capture by the rhodopsin molecules inside the fibre, but it can also be absorbed by external structures such as screening pigment granules, or even by adjacent receptors. When this happens there is 'crosstalk' between receptors, and resolution suffers. The practical consequence is that there is nothing to be gained by having receptors narrower than 1 μm, and this in turn sets a lower limit to focal length, and hence the size, of an eye with a given resolution.

Photoreceptors are typically long and narrow (the photopigment-bearing outer segments of human rods are about 25 μm long and 1–2 μm wide, and contain about 10^8 rhodopsin molecules). The proportion of light that a receptor absorbs depends on its length. Typically, vertebrate photoreceptors made of discs absorb about 3 per cent of the incident light for every μm of their length, and invertebrate receptors made of microvilli absorb about 1 per cent per μm. To absorb 90 per cent of the light reaching it a vertebrate receptor would need to be 77 μm long, and an insect receptor 230 μm. These numbers are fairly typical of receptors in the two groups (human rods are rather short). The relationship between absorption and receptor length is logarithmic rather than linear because with increasing distance down the receptor there is less light left to absorb. The proportion of the incident light absorbed by a receptor of length L can be found from $(1 - e^{-kL})$ if the light is monochromatic (which is roughly true of the deep sea where only blue light penetrates), or from $[kL/(2.3 + kL)]$ for white light, typical of terrestrial conditions. k is the absorption coefficient – the proportion of the light absorbed per micrometre, if L is measured in micrometres. A useful discussion of receptor absorption, and in particular the way it depends on wavelength, can be found in Warrant and Nilsson (1998).

Resolution and eye design

We are now in a position to use the physical principles outlined in the preceding sections to draw some firm conclusions about the relationship between an eye's size and construction, and the resolution it provides. A satisfying way of doing this is to try to 'design' an eye to a particular specification. If this can be done, using these principles, we can be reasonably sure that nothing important has been left out.

Imagine a small vertebrate with a single-chambered lens eye similar in design to our own. This animal is a herbivore feeding in bright daylight (this avoids problems of photon scarcity, to be discussed in the next section). It needs to resolve grass at, say, 3 m, which approximates in angular terms to a 10 cycles per degree grating. Converting from degrees to radians gives a retinal sampling frequency v_s of 10×57.3, or 573 cycles per radian, and from eqn (3.2) this means that $f/(2s) = 573$. Waveguide considerations mean that the receptor separation s cannot be much less than about 2 μm (1 μm receptors and 1 μm spaces), so that the focal length f must be at least 573×4 μm, or 2.29 mm. One would expect that the retinal sampling frequency would match the optical cut-off frequency v_{co} quite closely, as in the human eye, avoiding either 'unused' resolution on the one hand or superfluous receptors on the other. v_{co} is thus also 573 cycles per radian, and from eqn (3.4) this means that $D/\lambda = 573$. If λ is 0.5 μm, it follows that the eye must have an aperture diameter D of 1.15 mm. Thus the main features of this fictitious eye – (its focal length, aperture diameter and receptor diameter) – all follow from the tasks that evolution has assigned to it, and the particular physical principles that apply to eyes.

Sensitivity

The consequences of low photon numbers

The statistical uncertainties associated with small photon numbers mean that at low light levels the potential resolution of an eye cannot be realised, as indicated in Fig. 3.1. The first clear demonstration that human rod receptors actually detect individual photons was made by Hecht, Schlaer and Pirenne in 1941, and Fig. 3.8, from Pirenne's book *Vision and the eye* (1967), is based on that study. All four figures show the image on the retina of the same bright field containing a dark circular patch, but at different levels of illumination. In I the light level is so low that only 6 of the 400 receptors receives a photon: this is approximately the situation at the human threshold of vision. (Although single receptors are capable of detecting single photons, the brain requires a 'safety factor' of about 6, so that responses are not made to spontaneous rhodopsin activations.) In II the light level is ten times higher, but still the dark patch is invisible, disguised in the 'noise' of the random background of photon hits. By III a gambler might be prepared to guess that that there was a dark region in the field, but it is only by IV, 1000 times the threshold level, that the dark patch stands out with certainty. This demonstration makes it clear just why it is that we see so badly in the dark.

At higher light levels than those illustrated in Fig. 3.8 it becomes possible to distinguish different shades of grey, and ultimately small contrasts such as occur in the images of gratings near the resolution limit (Fig. 3.3). To resolve these gratings, the ability to detect contrasts of a few per cent is essential. How much light,

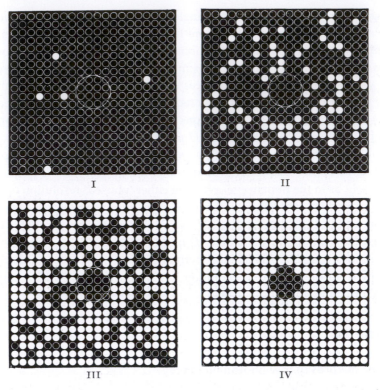

Fig. 3.8 Effect of low photon numbers. The four panels show a square of 400 receptors in the human retina with sample distributions of photon captures at the threshold of vision (I) and three higher light levels, with a factor of 10 increase between each (II–IV). The dark disc in the centre only becomes reliably detectable at intensities between 100 and 1000 times threshold. From Pirenne (1967).

or more precisely how many photons per receptor, are needed to detect a particular contrast? This question was first studied in the 1940s by Hugo deVries and Albert Rose, and the general answer they reached was that the minimum detectable contrast was proportional to the reciprocal of the square root of intensity (the Rose–deVries law, see Pelli 1990). In Box 3.2 we see how this rule arises from the statistics of photon capture, and then discuss its implications for vision.

It is impressive how big some of the photon numbers, predicted by Eqn (3.5), have to be. If the contrast in a grating is 0.5 (50 per cent) the number required is $1/0.5^2 = 4$. With a contrast of 10 per cent the number is 100, but when the contrast is down to 1 per cent, which humans can easily detect, the number is 10 000. These numbers refer to photons collected within one 'integration time' of the eye: roughly speaking, this is the time it takes for a receptor to respond fully to a change in intensity, and it is typically 0.1s or less. Thus at low contrast each receptor would require photon numbers around 10^5 per second. This is still an underestimate, because for a variety of reasons only a proportion of the photons

Box 3.2 How many photons are needed to detect a given contrast?

What has to be established is whether or not a real difference in intensity between two stripes in a grating, represented in the retina as a difference in photon numbers captured by the receptors, is larger than the 'noise' level, i.e. the statistical fluctuations in the numbers of photons arriving at the receptors. Fortunately, in this kind of statistics (Poisson distribution), noise and signal size are closely related. The variation in photon numbers, measured as the standard deviation, $\sigma(n)$, of repeated samples, is equal to the square root of the mean number, n, in the sample, i.e. $\sigma(n) = \sqrt{n}$. This property is common to many 'noisy' processes, for example current fluctuations in resistive circuits where small numbers of electrons are involved. Contrast (C) in a grating was defined earlier as the difference in intensity ($\Delta I = I_{max} - I_{min}$) between pairs of stripes, divided by the sum of the intensities, i.e. $C = \Delta I/2I$, where I is the average intensity. In a single sample pair we can replace intensities with photon numbers, which gives us $\Delta n/2n$ where n is the average photon number. For a brightness difference to be regarded as real, ordinary statistical reasoning suggests that the difference between the samples, Δn, should be greater than the standard deviation $\sigma(n)$, or, to give 95 per cent certainty, $2\sigma(n)$ (illustrated in Fig. 3.9). Thus a difference is detectable if $\Delta n > 2\sigma(n)$. To reach the answer, we need to do two things to this expression: divide both sides by $2n$, and replace $\sigma(n)$ by $\sqrt{n}$. This now gives $\Delta n/2n > 2\sqrt{n}/2n$. The left-hand side is, on average, equal to $\Delta I/2I$, which is the contrast C, and the right-hand side tidies up to simply $1/\sqrt{n}$. So the final result is:

$$C > 1/\sqrt{n}, \text{ or } n > 1/C^2. \tag{3.5}$$

The first expression is a version of the Rose–deVries law mentioned earlier, and the second tells us how many photons are needed, per receptor, to detect particular contrasts.

Fig. 3.9 Photon statistics and contrast detection. The figure shows the way photon samples will be distributed in receptors that image two areas of slightly different brightness. If the average difference in photon numbers (δn) is greater than twice the standard deviation of each distribution ($\sigma(n)$), the difference in brightness can be reliably detected.

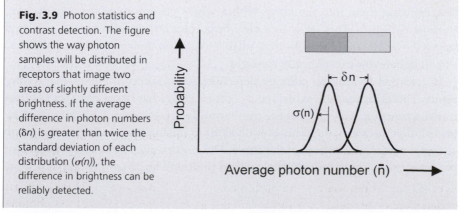

Fig. 3.10 Resolution and contrast loss. Low photon numbers limit the minimum detectable contrast. The effect of this is to set a 'floor value' to the contrast ratio in the contrast transfer function (see Fig. 3.3) which in turn limits the maximum detectable spatial frequency (ν_{max}) to a fraction of the cut-off frequency (ν_{co}). In this case raising the minimum contrast ratio to 0.32 reduces the maximum frequency to about 58 per cent of the bright light value.

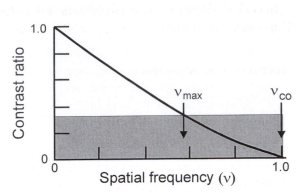

reaching the eye from a scene are actually absorbed by rhodopsin molecules. In humans this is around 10 per cent, which means that the photon numbers needed to detect a 1 per cent contrast are close to a million per second per receptor. This is a very large number, and the obvious next question is: 'How many photons does the world provide for us to see with?'

Available photon numbers

The radiance (R) of a white card in bright sunlight (a measure of the number of photons it emits), is about 10^{20} m^{-2}.sr^{-1}.s^{-1}, in room light it is about 10^{17}, in moonlight 10^{14}, and in starlight 10^{10} – the absolute threshold for human vision. (The meaning of the units is explained in Chapter 2.) These numbers seem enormous, but they reduce by a factor of 10^{12} in going from square metres in outside space to the dimensions of photoreceptors which are measured in square micrometres. Similarly the cones of light accepted by single receptors are typically less than 1 square degree, and as there are 3283 square degrees in a steradian (which is a cone 65.5° across), this reduces photon numbers by a further 10^3 or more. This cuts down the final numbers available to receptors to a million per second or less, bringing them into the range within which photon numbers start to limit contrast detection. This leads to a very important conclusion: eyes are 'photon starved' – (in the sense that they are unable to exploit their potential capabilities) – at all light levels except bright daylight.

In addition to limiting contrast detection, low photon numbers also reduce acuity. This is most easily explained by considering what happens to the contrast transfer function (Fig. 3.10). If the minimum detectable contrast is increased by low photon numbers, then this is equivalent to raising the baseline of the graph so that it cuts off the bottom of the curve. Thus with only 10 photons per receptor per integration time available, the contrast limit will be 32 per cent (according to eqn 3.5) and that will have the effect of limiting the maximum detectable spatial frequency to less than 0.6 of the value in bright light, i.e. the cut-off frequency. This is basically why fine work requires high light levels.

Making eyes more sensitive

From what has been said, it is clear that the more photons an eye can capture the better. This is important at normal light levels, but the pressures are even greater for nocturnal animals (moonlight is a million times dimmer than sunlight), and those that live at depth in the ocean, where even in the clearest water light is reduced by a factor of 10 for every 70 m. We can call this ability to capture photons an eye's *sensitivity*, and define it as the number of photons (n) caught per receptor when the eye views a scene of standard radiance R.

There are basically two features that make an eye sensitive: these are the pupil diameter D and the angle in space over which each receptor accepts light ($\Delta\rho$). For present purposes $\Delta\rho$ is given by d/f, the angle the receptor's diameter makes at the eye's nodal point (see Fig. 3.11). Receptor length can be important too, as discussed in the earlier section on Photoreceptor Optics, and the term P_{abs} is added here to take into account the proportion of photons entering the receptor that are absorbed by the photopigment (usually between 0.1 and 0.9; see Photoreceptor optics, above). The sensitivity S is then given by:

$$S = n/R = 0.62\, D^2\, \Delta\rho^2\, P_{abs}. \tag{3.6}$$

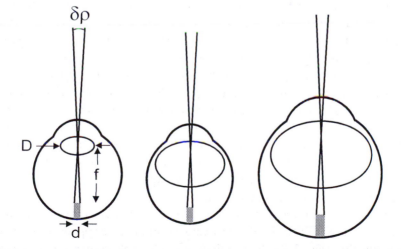

Fig. 3.11 Increasing an eye's sensitivity. The sensitivity of the eye on the left can first be improved by widening the aperture (*D*) to the maximum possible (centre). After that, the only way to increase photon capture without changing resolution (constant acceptance angle $\Delta\rho = d/f$ radians) is to scale up all three parameters (*D*, *d*, and *f*) together.

This equation is quite easily derived from photometry, and a full explanation can be found in Land (1981). (The factor of 0.62 is $(\pi/4)^2$, and arises because both aperture and receptors have circular cross-sections. Note that for small angles $\Delta\rho^2$ is a solid angle in steradians). What is important here is that there is really only one variable that can safely be varied to improve sensitivity, and that is the aperture D. Increasing $\Delta\rho$ will also increase sensitivity, but at the expense of resolution since in a single chambered eye $\Delta\rho$ and the inter-receptor (sampling) angle $\Delta\phi$ are almost the same (Fig. 3.2).

Let us look at how, in practice, an eye might become more sensitive (Fig. 3.11). Initially the aperture could be increased: in going from a daylight diameter of 2 mm to 8 mm at night, the human pupil increases the eye's sensitivity by a factor of 16. In really nocturnal animals such as owls and opossums the pupil is almost as wide as the eye itself. Obviously, however, D cannot be greater than the eye diameter, so ultimately it must be eye size that limits sensitivity. Any further sensitivity increase requires a larger eye to accommodate the larger pupil, and indeed most nocturnal animals have large eyes. However, increasing the eye size, and hence the focal length f, will actually decrease $\Delta\rho$, which is given by d/f, so to reap the rewards of the larger eye the receptors must be made correspondingly wider.

Table 3.2 The sensitivity (S) of a selection of animal eyes

Name	Sensitivity	Light habitat	Ref
Cirolana (marine isopod)	4200	deep sea	1
Oplophorus (decapod shrimp)	3300	deep sea	2
Dinopis (ogre-faced spider)	101	nocturnal	2*
Limulus (horseshoe crab)	83–317	coastal mainly nocturnal	1*
Ephestia (moth)	38	nocturnal/crepuscular	2*
Onitis aygulus (dung-beetle)	31	nocturnal/crepuscular	3*
Phronima (hyperiid amphipod)	38–120	mid-water	1
Man (peripheral rod pool)	18	crepuscular	2*
Pecten (scallop)	4.0	coastal sea-floor	2*
Bufo (toad)	4.0	mainly diurnal	4*
Leptograpsus (shore crab)	0.5	diurnal	1*
Onitis ion (dung beetle)	0.35	diurnal	3*
Worker bee	0.32	diurnal	2*
Phidippus (jumping spider)	0.04	diurnal	2*
Man (fovea in daylight)	0.01	diurnal	2*

References to original data: 1. Land and Nilsson (1990) 2. Land (1981) 3. McIntyre and Caveney (1998) 4. Warrant and Nilsson (1998). * Values recalculated for white light using method given in 4.

The monochromatic light formula $S = 0.62\, D^2\, \Delta\rho^2\, (1 - e^{-kL})$ was used for the three deep-sea species (no *), for all the others (*) the white light formula $S = 0.62\, D^2\, \Delta\rho^2\, (kL/(2.3 + kL))$ was used. $\Delta\rho$ is obtained from d/f, the receptor diameter divided by the focal length.

In vertebrates' eyes the receptors themselves are not particularly large in big eyes. What tends to happen instead is that small receptors are grouped into larger units at the level of the retinal ganglion cells, so that the *effective* receptor diameter is increased (spatial summation). This arrangement is often quite flexible, so that the size of the receptor 'pool' can vary with light level, allowing a trade-off between high resolution in daylight (small effective $\Delta\rho$) and high sensitivity at night (large effective $\Delta\rho$). Another strategy is to collect photons over a longer period of time (temporal summation). As with spatial summation there is a penalty, in this case the increased movement blur that results from the lengthened 'shutter time'. Nevertheless, when used appropriately, spatial and temporal summation can be very effective in augmenting the purely optical adaptations summarized in Eqn (3.6). For example, it has been estimated that, with optimal summation, the locust eye can extend its visual range down to light intensities 100 000 times dimmer than that provided by the optics alone (Warrant 1999).

Even without taking spatial and temporal summation into account, the range of sensitivities that different eyes obtain by varying the parameters in eqn (3.6) is remarkably large. For the human eye in daylight ($D = 2000$ μm, $\Delta\rho = 1.2$ 10^{-4} rad, $P_{abs} = 0.31$) S is 0.01 μm^2.sr, whereas at the other extreme the deep-sea isopod crustacean *Cirolana* ($D = 150$ μm, $\Delta\rho = 0.78$ rad, $P_{abs} = 0.51$) has a value for S of 4200 μm^2.sr. If both eyes were looking at the same scene, the crustacean would capture 420 000 times as many photons per receptor, a ratio not far short of the million-fold difference between daylight and moonlight, although well short of the total range of usable human vision (10^{10}). Sensitivity figures for a range of animals are given in Table 3.2. In general there is excellent agreement between the value of S, and the light regime in the animal's habitat. Diurnal and surface-living animals tend to have S-values below 1, for crepuscular and mid-water animals S is in the range 1 to 100, and for nocturnal and deep-water animals it is between 100 and 10 000. Another selection of sensitivity values, showing the same trend, is given by Warrant and McIntyre (1990).

Conclusions

A 'good' eye can be defined as one that resolves well under a variety of lighting conditions, and we are now in a position to see what anatomical features make this possible. The first point is that such an eye will have to be reasonably large, for three reasons. A long focal length is needed to obtain a low minimum resolvable angle and a high retinal sampling frequency (eqn 3.2); a wide aperture is needed to reduce diffraction, and thus ensure a high optical cut-off frequency (eqn 3.4); and a wide aperture is also needed to get enough light into the eye to ensure adequate photon numbers, and thus good contrast detection in dim light (eqn 3.5 and 3.6). Large absolute eye size benefits both resolution and sensitivity, so it is no surprise to find an evolutionary trend towards large eyes in all animals that require good eyesight. Humans, hawks, and dragonflies have large eyes in order to resolve well, whereas cats, owls, and moths use eye size more to improve sensitivity. Not surprisingly, hunters in the deep sea, requiring both resolution and sensitivity, sometimes have huge eyes. The largest recorded eye is that of a deep-sea squid, and it had a diameter of 40 cm. Conversely, low-acuity eyes operating in daylight can be less than a millimetre wide.

The differences between diurnal and nocturnal eyes are mainly in the size of the aperture and the angle in space ($\Delta\rho$) over which each receptor accepts light. Human eyes have relatively small pupils, with an F-number (f/D as in photography) between 8 in daylight and 2 at night. Diurnal insects such as bees typically have F-numbers of about 2. In fishes and other nocturnal vertebrates the F-number is closer to 1, and in some arthropods such as moths and lobsters it can be as low as 0.5. Since image brightness varies as $(1/F\text{-number})^2$, this means that the optics of a lobster eye at night have 256 times the light-catching power of a human eye in daylight. There are no advantages for a diurnal eye in having a pupil larger than is needed to prevent diffraction from limiting image quality. Indeed, there are disadvantages, because other defects such as spherical and chromatic aberration become worse. But in the dark high resolution becomes unusable, and the need for photons is paramount, which dictates that the aperture should be as large as possible.

In daylight there are plenty of photons, and the narrower the receptors are the better, because this means that the eye can have a short focal length for a given resolution (eqn 3.2), and so be physically small. Because there is a lower limit of about 1 μm to the receptor diameter, imposed by waveguide optics (Fig. 3.7), this means that focal lengths cannot become vanishingly small either. There is thus a minimum size an eye must have, however bright the light. In the dark wider receptors are favoured, because this increases the angle ($\Delta\rho$) over which they capture photons. This means, that in an eye of a fixed size resolution will theoretically be reduced if sensitivity is increased. However, in dim conditions fine resolution is anyway unusable (Fig. 3.10), so the compromise between resolution and sensitivity favours wider receptors. There seems to be a practical upper limit to receptor diameter of about 25 μm; this is found in lobsters and some other crustaceans. Such receptors will accept 100 times more photons than a 2.5 μm cone from the human fovea. The clever way of managing this trade-off between resolution and sensitivity, so that the eye has the best resolution available to it at different light levels, is to have small receptors, but to pool them into larger assemblages in darker conditions. There is a good deal of evidence to suggest that this is what occurs, certainly in the eyes of vertebrates.

Summary

1 Eyes can be characterized by their resolution and sensitivity. Resolution is the fineness, in angular terms, with which the optical environment is sampled. Sensitivity is quantifiable as the number of photons a receptor receives when the eye is viewing a scene of standard luminance.

2 Resolution depends on the sampling density of the retinal receptors and also on the quality of the optical image. This quality can be affected by defects of focus, and by spherical and chromatic aberration. It is ultimately limited by diffraction (interference of light waves in the image). The larger the aperture of an eye, the smaller the effect of diffraction.

3 Because of waveguide effects photoreceptors cannot be made narrower than 1–2 μm without compromising resolution.

4 In dim light the ability to detect contrast is limited by the numbers of photons that receptors can obtain. The smaller the number of photons caught the worse the statistical quality of the image. Photon numbers are maximized in high sensitivity eyes by the use of high relative apertures (aperture diameter/focal length) and wide receptors.

4 | Aquatic eyes: the evolution of the lens

Evolutionary origins

Life began in the sea, and we, as land-living animals, have features of our eyes that reflect that watery ancestry. In particular we have a lens with a peculiarly inhomogeneous structure, which supplements the ray-bending power of the cornea. In terrestrial animals it is the curved air–fluid interface of the cornea that performs most of the optical work of bringing light to a focus, but in aquatic animals this surface has no optical function. It exists of course, but with fluid on both sides it has no capacity to refract light. Thus, with rare exceptions, the lens is the only optical structure capable of producing an image in water.

The vertebrate fossil record has, unfortunately, almost nothing to say about the origins of the eyes or their optical systems. The lampreys, relatives of the jawless ostracoderm fishes of 450 million years ago, have eyes that are, for all practical purposes, the same as those of other modern fishes (Nicol 1989). Other modern relatives of the earlier chordates such as *Amphioxus* have pigmented photoreceptors, but nothing resembling a real eye. However, amongst the molluscs it is possible to make out a series of eyes of modern forms that at least provides a clue to the early evolution of eyes of the single chambered type. Figure 4.1 shows such a series. In the limpet *Patella* the eye is a V-shaped pigmented pit containing receptors. Each receptor has an acceptance angle of 90° or more, restricted only by the shadowing effect of the pigment behind it. Pit eyes like this are common throughout the 'lower phyla', and enable an animal to locate lighter or darker regions of the environment. In many gastropods, the abalone *Haliotis* for example, the mouth of the pit is drawn in to give the eye a more spherical shape, and a narrower opening, restricting the acceptance angle of each receptor to perhaps 10°. While this results in an improvement in the eye's resolution, it is obvious that to pursue this line any further will produce eyes in which less and less light reaches the image. Thus this is not a particularly good evolutionary route to follow. The only animal to have pursued this route to its logical conclusion is the ancient cephalopod mollusc *Nautilus*. A much better solution is to evolve a lens. In the snail *Helix* this is simply a ball of jelly which

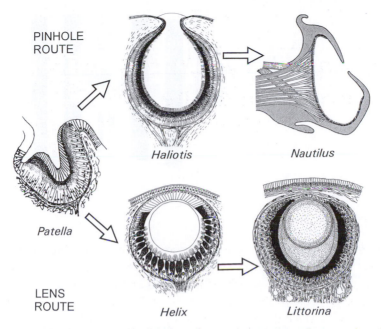

PINHOLE
ROUTE

Haliotis   *Nautilus*

Patella

LENS
ROUTE   *Helix*   *Littorina*

Fig. 4.1 Examples showing two possible directions of eye evolution in the molluscs, starting with the pigmented pit eye of the limpet, *Patella*. *Top row*: pinhole eyes in the abalone *Haliotis* and the cephalopod *Nautilus*. *Lower row*: lens eyes in the land snail *Helix* and the shore-living gastropod *Littorina* (*Patella*, *Haliotis* and *Helix* from Hesse 1908, *Nautilus* from Young 1964, Littorina from *Newell* 1965). The Nautilus eye is about 10 mm across, the others are all less than 1 mm.

converges the light rays a little, though not enough to form a sharp image (Fig. 4.1). However, in the periwinkle *Littorina*, and many other gastropod molluscs, the lens has evolved into a sophisticated structure with a graded refractive index, and excellent image-forming capabilities.

The pin-hole eye of Nautilus

What distinguishes the *Nautilus* eye from other lensless eyes is its size (Fig. 4.1). Most of the lensless eyes we have mentioned so far are a fraction of a millimetre in diameter, with a few hundred receptors. In *Nautilus*, however, the eyes are nearly a centimetre in diameter, comparable in size with the lens-containing eyes of *Octopus*, or indeed of many fish. Thus these are serious eyes, and this impression is reinforced by the discovery that the pin-hole pupil can vary its diameter with light intensity, between 0.4 and 2.8 mm. The eye is also equipped with a series of muscles that rotate it in such a way as to stabilize its vertical axis against the rocking motion of the animal as it swims. We know little about the functions of vision in *Nautilus*, but the animal does have an optomotor response (Fig. 4.2); that is, it can be made to rotate or swim in circles by rotating a striped drum

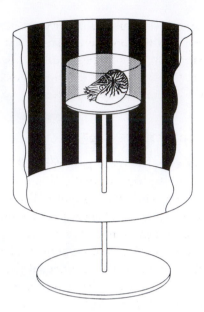

Fig. 4.2 *Nautilus* in an optomotor apparatus that elicits circular swimming when the stripes are rotated. The dish containing the animal is 25 cm across. From Muntz and Raj (1984).

around it (Muntz and Raj 1984). The finest stripe pattern that will produce this response subtends between 11 and 22°, which is roughly what one would predict with a partly open pin-hole. The real function of this behaviour is to prevent the eye or body from rotating when a *stationary* background is present, and it may be that it serves to stabilize the swimming of the animal as it browses along the reef.

The weakness of the *Nautilus* eye is that the pin-hole arrangement is a very unhappy compromise. To improve the resolution to anywhere near the levels of a lens eye means decreasing the size of the pupil to a diameter that hardly lets in any light at all. With a 0.4 mm pupil the angle in space over which a single receptor accepts light is about 2.3°, roughly comparable with the resolution of the eye of a small insect. However, the image is then dimmer than the image in a fish eye by a factor of about 400. Since opening up the iris results in a disastrous loss of resolution, the animal is trapped in a visual world that is by most standards unusably dim or unusably blurred.

The way out of this is to evolve a lens which sharpens the image by focusing rather than by shading. The real enigma of *Nautilus* is that it has not managed this, in spite of having had nearly 500 million years to do so. Other cephalopods (octopus, squid, and cuttlefish) have excellent lenses very much like those of fish, and numerous gastropods, worms and arthropods have also managed this feat. It should not be difficult, because it can proceed in small steps, all of them representing an improvement (see Chapter 1). Thus a small blob of jelly or mucus, with a refractive index somewhat higher than the surrounding water, placed in the pupil of the eye, will converge entering rays slightly, and this in turn will reduce the width of the blur circle on the retina without requiring a decrease in pupil diameter. The process of improvement could continue until the 'lens' converged the light to a point on the retina, at which stage the transformation to an image-forming eye would be complete. Using conservative assumptions Nilsson and Pelger (1994) estimated that this whole evolutionary process would take less than half a million years to complete.

Spherical lenses

Producing a lens that will perform well in water is not quite as easy as it may seem at first. It turns out that a lens made simply of a glass-like material (dry crystalline protein for example) will not produce an image of good enough quality, nor have a focal length short enough to be really useful. The focal length needs to be kept short in relation to the size of the lens to keep the eye as a whole reasonably small. This means that the radii of curvature of the surfaces have to be small, which in turn makes a spherical shape for the lens more or less obligatory. However, spherical lenses have serious defects. The worst is known as spherical aberration (Fig. 3.6), in which rays at a distance from the axis of the lens are bent through too great an angle to come to the same focus as the on-axis rays, and the result is a blur circle on the retina rather than a sharp image (Fig. 4.3a). This would be wide, with a spherical lens, and the image would be very poor. The

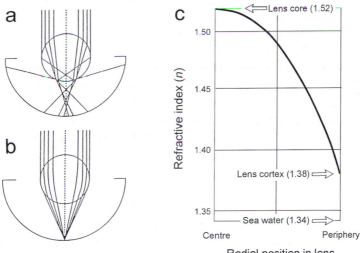

Fig. 4.3 (a) Paths of rays through a homogeneous lens of refractive index 1.66, showing how rays far from the axis are refracted too much (spherical aberration). (b) A lens with the same focal length as (a), but with a gradient of refractive index, and a maximum index of 1.52 in the centre. Note that rays are bent continuously and come to a common focus. (c) Form of the gradient in a fish lens, capable of producing an image free from spherical aberration. (a) and (b) from Pumphrey (1961); (c) based on Jagger (1992).

other problem is that the lens would have a rather long focal length. A single surface of radius r, separating two media of refractive indices n_1 and n_2, has a focal length given by $rn_2/(n_2 - n_1)$. A spherical lens, where light encounters two surfaces, both of radius r, has a focal length (f) equal to half this:

$$f = 0.5rn_2/(n_2 - n_1). \tag{4.1}$$

The refractive index (n_2) of a dry protein such as the crystallin found in lenses is about 1.53, and with sea-water ($n_1 = 1.34$) as the outside medium, the focal length of a lens made of such material would be 4 lens radii. In fact, this is much longer than the focal lengths of real lenses in fish and cephalopods. It has been known since the studies of Matthiessen in the 1880s that the lenses of fish as well as cephalopods and marine mammals nearly all have focal lengths of about 2.5 lens radii, a number that has become known as Matthiessen's ratio.

Lenses with refractive index gradients

Clearly, a spherical lens made of homogeneous protein does not fit with what we know of fish lenses, namely that they are of excellent optical quality and short focal length. This apparent contradiction interested a number of nineteenth-century scientists including James Clerk Maxwell, who came up with the idea that such lenses must have a gradient of refractive index highest in the centre and lowest near the periphery. Matthiessen had shown that there was such a gradient in fish lenses, and believed that its form was that of an inverted parabola, with the refractive index falling as the square of the distance from the lens centre (Fig. 4.3c). Matthiessen, it turns out, was not far wrong in his guess, although more recent theoretical studies have suggested that there are other functions that give a somewhat better performance in terms of the correction for spherical aberration (for a review see Jagger 1992).

What does the refractive index gradient achieve? In the first place it changes the pattern of refraction from a discrete bending of the rays at each interface to one in which rays are bent continuously within the body of the lens. The effect on spherical aberration is that the outermost rays, which travel shorter distances within the lens, are bent relatively less than they are at the interfaces of the homogeneous lens (Fig. 4.3b). Given the correct gradient, all rays can be brought to a focus at the same point, for light of a single wavelength. The shorter focal length is achieved because continuous refraction results in greater total ray-bending than does 2-surface refraction. In fact, an f/r ratio of 2.5 can be achieved in a gradient index lens with a central refractive index of 1.52, whereas the same ratio would require a homogeneous lens to have an index of 1.66. The real value of the short focal length of fish lenses lies in the effect this has on light-gathering power. In photographic terms the F-number of the eye (focal length/diameter) is

1.25, which gives an image 2.6-times brighter than the image behind a homo-geneous protein lens ($n = 1.52$) with an F-number of about 2.

The other important defect of biological lenses in general is chromatic aberra-tion, in which light of shorter wavelengths is brought to a focus closer to the lens than longer wavelength light (Fig. 3.6). This means that a single retina at a fixed distance from the lens cannot be in focus for all wavelengths simultaneously. For animals that have only one visual pigment (deep-sea fish and most cephalopod molluscs, for example) this is not a problem. However, shallow water fish have excellent colour vision, and typically they possess 4 cone types whose wave-lengths of maximum sensitivity cover a 250 nm range from ultraviolet to red. One way round the problem would be to place the different cone types at differ-ent distances from the lens, and there is some evidence for this. However, the distances involved are quite large (up to 10 per cent of the average focal length, or 1 mm in a 10 mm focal length eye) and cone separations as great as this are not physically possible in a thin retinal sheet. It is now clear that some fish use another method. This is to produce lenses with multiple focal lengths, brought about by variations in the basic Matthiessen gradient (Kröger *et al.* 1999). The way this works is shown in Fig. 4.4. For light of a single wavelength, the inner zones of the lens bring light to a closer focus than the outer zones (this effectively means that the lens is *over*-corrected for spherical aberration). However, for white light with a range of wavelengths the images from inner and outer zones have a spread of focal lengths, because of chromatic aberration. This means that the position of the image for short wavelengths formed by the outer zone can be made to coincide with the image for long wavelengths formed by the inner zone. This in turn allows cones with different wavelengths of maximum sensitivity to

Fig. 4.4 Method used by some fish to overcome chromatic aberra-tion. By slightly varying the refrac-tive index gradient (Fig. 4.3c) the lens produces several sharp images at different distances. Although each of these images suffers from chromatic aberration, their loca-tions can be adjusted so that the images for different wavelengths coincide as shown. This means that cones with different spectral sensitivities can be arranged in a single layer, and each receive a sharp image. F_1 and F_2, foci from different lens regions. B and R, foci for blue and red light. Based on Kröger *et al.* (1999).

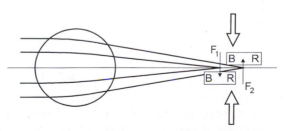

Plane of sharp focus
for blue and red rays

receive in-focus images in the same plane. This is not a perfect solution, because these in-focus images are contaminated by light from the other out-of-focus images, and so will have reduced contrast compared with a perfectly corrected monochromatic image. However, this is better than not having a sharp image, and it seems that a wide variety of teleost fish have opted for this solution, and possibly other vertebrates too. Figure 4.4 is somewhat over-simplified; in the species studied by Kröger *et al.* (1999) there were three distinct images corresponding to the sensitivity maxima of the three cone types, rather than the two shown in the figure.

The 'Matthiessen lens' is a winning design, not only in terms of its optical performance, but also its evolutionary popularity. Because an f/r ratio of around 2.5 immediately tells one that a spherical lens has a gradient structure, it is easy to survey the animal kingdom for occasions on which this type of eye has evolved. It evolved in the fish, presumably once, in the cephalopod molluscs (Figs 4.5b, 4.6), and possibly more than once in the gastropod molluscs as it is found in pulmonates such as the fresh-water snail *Lymnea*, as well as a variety of prosobranchs. These include the remarkable eyes of heteropod molluscs (*Pterotrachea, Oxygyrus*) with large spherical lenses and long, narrow scanning retinae (Fig. 4.5a). It also evolved independently in two unlikely places: among the

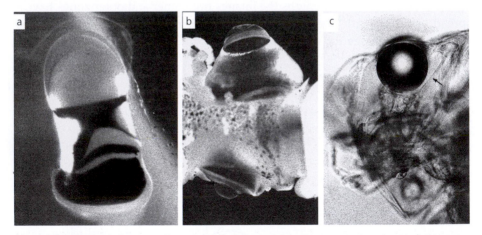

Fig. 4.5 Photographs of spherical lens eyes in molluscs and a crustacean. (a) Eye of *Pterotrachea*, a carnivorous sea-snail (Heteropoda). The eye is 3 mm long, with a long narrow retina only six cell rows wide. (b) Unusual eyes of a mid-water squid (*Histioteuthis*). The two eyes are different sizes. The larger 'telescopic' eye has a yellow lens and is directed towards the sea surface, the smaller eye has a clear lens, a wider field of view, and is directed downwards. Partly dissected. The upper eye has a diameter of 9 mm. (c) The pontellid copepod *Labidocera* is probably unique among crustaceans in possessing a pair of eyes with Matthiessen's ratio lenses. The top of the small eye-cup containing the retina is visible beneath the lens, and to the right of the lens is a thin striated muscle (*arrowed*) which moves the eye-cup (see Chapter 9). At the bottom of the picture is the third (ventral) eye, which has a quite different construction. Lens diameter 145 μm.

annelid worms in the alciopids, a family of polychaetes that have become active carnivores in the marine plankton; and just once in the Crustacea, in the copepod *Labidocera* (Fig. 4.5c) where the males have a pair of lenses that share a line-like retina of 10 receptors (see Chapter 9). Remarkably, there are also spherical lenses in the cnidarians, in the curious inward-pointing eyes of cubomedusan jellyfish; again, these have a refractive index gradient and a focal length close to 2.5 lens radii. So far, there are no known cases of image-forming spherical lenses with long focal lengths and homogeneous internal structures, except in some gastropods such as *Helix* (Fig. 4.1) and some polychaete worms. Here the 'lens' is not much more than a jelly-like mass filling the cavity of the eye, and optically such an eye is closer to a pigment cup than a camera-type eye. It seems that if an eye starts out on the evolutionary path that leads to a spherical lens it nearly always goes all the way, evolving a lens with a gradient of refractive index which produces an almost aberration-free image.

Eyes of fish and cephalopods

The similarities between these two groups of swimming animals provide some of the best known examples of convergent evolution. Of these, the structure of the eye is perhaps the most astonishing (Fig. 4.6). Packard (1972) put it like this:

> … the modern cephalopod eye, with its single chamber, lens, ciliary body, iris, hemispherical retina, cartilaginous sclera and external argentea is also the most clamorously vertebrate-like structure of the cephalopod organization.

Because cephalopods and vertebrates have very separate evolutionary origins, we can be certain that the similarities exist because both groups have hit upon, and perfected, the same engineering solution to the problem of seeing well in the marine environment.

There are crucial differences, however. The retina in cephalopods has a structure in which the photoreceptors have their photopigment-bearing regions directed forwards, towards the light, whereas in vertebrates they are (for some quirk of development) situated at the back of the eye with the photosensitive region pointing away from the light (Fig. 4.6). It is often remarked that the vertebrate retina is the wrong way round, and this is true, but because the overlying cells and nerve fibres in the vertebrate retina are reasonably transparent, the optical handicap is small. A great deal of processing occurs in the vertebrate retina, with its three sequential layers of nerve cells (the receptors, bipolars and ganglion cells) separated by the two layers of laterally extending neurons – the horizontal and amacrine cells (a good account of the structure and function of the fish retina can be found in Nicol 1989). In the cephalopod retina there is no such layered arrangement, and most of the processing occurs outside the eye, in the

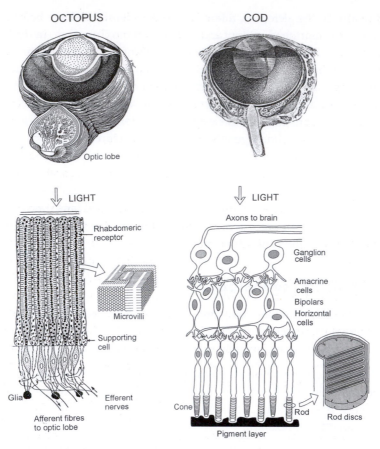

Fig. 4.6 Convergence between the eyes of cephalopod molluscs and fish. The overall structure of the eyes is very similar (top; *Octopus* from Young, 1964, cod (*Gadus*) from an engraving by D.W. Soemmerring, 1818). Both eyes are large, 10 mm or more in diameter, and have spherical lenses whose centres are about 2.5 lens radii from the retina. The lower figures show that the retinae are completely different. In cephalopods the receptors are 'rhabdomeric', i.e. they are composed of photopigment-containing microvilli. Very little neural computation is done in the retina; this occurs in the optic lobe behind the eye. In the vertebrate retina (*right*) the rods and cones have receptive regions (the outer segments) composed of discs which carry the photopigment. In front of the rods and cones (with respect to the light path) are two layers of neurons in series (the bipolars and ganglion cells) with horizontal cells and amacrine cells forming lateral connections between each layer. Note that the receptors in *Octopus* point towards the light, but the vertebrate receptors point away. Octopus from Young (1964), vertebrate retina from various sources.

optic lobe of the brain. The receptors themselves are different, too, with the photo-pigment carried on microvilli in cephalopods (this is the typical arrangement for most invertebrates), but contained in disc-like structures in the rods and cones of fish. Cephalopods generally have only one visual pigment and are colour-blind

(Messenger 1991). The exception is the Japanese firefly squid, *Watasenia scintillans* which has three pigments based on different chromophore groups rather than different opsins (see Chapter 2). Most fish, on the other hand, have a range of visual pigments in their cones extending from the ultraviolet through to the red region of the spectrum, and excellent colour vision.

The lenses in the two groups are very similar in their optical properties, but they are not constructed the same way. The fish lens is a single structure surrounded by living cells, but cephalopod lenses develop in two parts, with the front and rear regions separated by a sheet of live cells. Remarkably, both parts have similar refractive index gradients, as required in a lens well corrected for spherical aberration.

In both fish and cephalopods there are muscles associated with the eye. External eye muscles are concerned with moving the eye in its orbit, and internal muscles focus the lens and adjust the iris. It seems astonishing that similar arrangements should be found in the two groups, but again the reason seems to be that the design simply requires them. Large eyes have to be stabilized, or motion blur will wreck the excellent resolution obtained by having a fine-grain retina and a long focal length. Similarly, depth of focus becomes smaller as eyes get bigger, making focusing mechanisms essential. In fish, as in other vertebrates, six muscles move the eyes – one pair for each axis of rotation. In the cephalopods the pattern of the muculature is less obvious; 13 muscles have been described in the cuttlefish *Sepia*, although this may boil down to six functional groups. In *Octopus* this six-group structure is more obvious. Like the semi-circular canal system of vertebrates, the statocyst in *Octopus* provides information about the animal's own rotation, and this, together with information about image motion from the eye itself, is used to counter-rotate the eye as the animal turns. The effect is to keep the image more or less still on the retina, the body effectively rotating around the stationary eye. However, the eye has to move from time to time, and in both cephalopods and vertebrates this is achieved by a fast flick-like movement known as a saccade, during which the eye is effectively blind. In both cases the strategy seems to be to minimize the time that the eye is moving relative to the surroundings, with consequent blurring of the image (see Chapter 9).

In fish there is a variety of focusing mechanisms (see Chapter 5, Fig. 5.9). Lampreys and bony fishes have a 'negative' accommodating mechanism in which the resting eye is focused for near objects, and muscular action shifts the focus to more distant objects by moving the lens back towards the retina. In cartilaginous fishes and amphibians muscle action has the opposite effect, 'positive' accommodation moving the lens away from the retina and so bringing nearer objects into focus (Walls 1942). Cephalopods seem to have both types of mechanism, different sets of muscles moving the lens in either direction, but how this arrangement works in practice is still not clear (Messenger 1981).

Matching eye to environment

Most eyes with spherical lenses look remarkably similar, mainly because Matthiessen's ratio fixes the proportions of the lens and eye-cup. However, they certainly differ in size if not in shape. The smallest and largest both belong to molluscs. The pond snail *Lymnea*, which has a perfect Matthiessen lens, is only about 0.5 mm across, whereas the largest giant squid for which there is reliable information had an eye 40 cm across – the size of a television set. The largest fish eyes are found in swordfish and tuna, where they attain a diameter of about 10 cm. Larger eyes still, 20–30 cm in diameter, were present in deep-diving ichthyosaurs which died out about 90 million years ago (Montani *et al.*, 1999). Large size can buy either high acuity or high light-gathering power, and it seems here that it must be the latter, because no fish exploits resolution anywhere near the diffraction limit of the lens. Eye size probably determines the extent to which fish can hunt into the night, or to what depths they can usefully operate. In the case of the giant squid one could argue that catching prey at the end of tentacles several metres long, deep in the ocean, requires both good resolution as well as high sensitivity.

Adaptation to the nature of the fish's environment is seen most clearly in the retina, and specifically in the way the ganglion cells are distributed. The output from the eye consists of the fibres of the optic nerve, which are the axons of the ganglion cells. The requirements of flexibility and economy of space limit the number of optic nerve fibres to about a million (compared with something like a

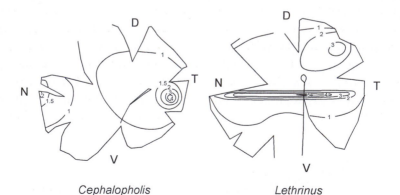

Cephalopholis   *Lethrinus*

Fig. 4.7 Patterns of ganglion cell density, reflecting the nature of the visual habitat, in the retinae of two fish. *Cephalopholis*, which lives in crevices in coral reefs, has a small forward-pointing 'area' of high density (see also Plate 1). *Lethrinus*, which swims over sand, has a pronounced horizontal 'visual streak' corresponding to the lateral view of the bottom. The ganglion cells are the outputs from the retina (see Fig. 4.6) and their distribution gives a good indication of an animal's visual priorities. Numbers are densities in thousands per square millimetre. N, T, D, V are nasal, temporal, dorsal, and ventral regions of the retina. Note that the lens inverts these relations, so that the temporal retina at the rear of the eye views the forward direction. Based on Collin and Pettigrew (1988).

hundred times as many photoreceptors), making this a real bottleneck in the visual pathway. This is why there is need for economy in the ganglion cell distribution, with the greatest numbers associated with parts of the image where there is the most information. Figure 4.7 shows the ganglion cell distributions in the dissected out retinae of two fish from different habitats. One of these, *Cephalopholis miniatus*, lurks in crevices in the coral reef, and the other *Lethrinus chrysostomas* lives over an open sandy bottom. *Cephalopholis* has a small region of high ganglion cell density in the temporal, forward pointing, region of the retina. To provide a forward field of view the lens has an 'aphakic' space in front of it (Plate 1). In contrast, *Lethrinus* has a long high density 'visual streak' which reflects the fact that most of the interest in this fish's world lies close to the horizontal plane. A similar streak is found in the retinae of grassland mammals such as rabbits, and sea-birds where the surface of the ocean only occupies a narrow horizontal strip in the visual field. One particularly interesting specialization of this kind occurs in the surface-feeding fish *Aplocheilus lineatus* which has two parallel visual streaks separated by about 40°. Apparently this gives the fish two views of its prey, which might be a drowning insect. One streak views the water surface from below, while the other looks out of the water just above the edge of 'Snell's window', the horizon for refracted rays, and thus sees the upper part of the prey (Fig. 4.8).

The deep sea, below a few hundred metres, provides a rather special environment. Little light penetrates from the ocean surface, and what does is predominantly blue, and limited to a relatively narrow downward-pointing cone (Lythgoe 1979). Nevertheless, many fishes still use this light to hunt by, presumably sighting potential prey by the dark silhouette cast against the dim residual skylight. (The evidence for this comes chiefly from the care that mid-water fishes and crustaceans take to disguise their silhouette with downward-pointing photophores, which emit light at an intensity adjusted to the downwelling light.) Where photons are scarce, the detectability of prey will depend on the amount of light reaching the predator's retina, and this puts a premium on the size of the lens. Big lenses imply big eyes, but many predatory mid-water fish have

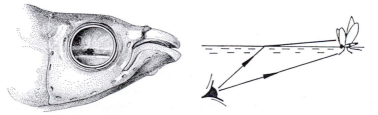

Fig. 4.8 *Left*: head of the surface-feeding fish *Aplocheilus* with lens of the eye removed, showing the two visual streaks on the retina. From Munk (1970). *Right*: illustrating how an object in the surface film can be viewed both above and below the surface.

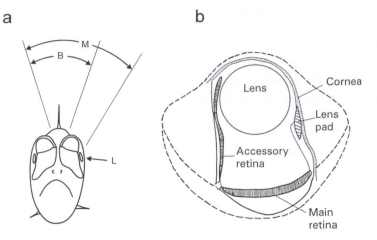

a

b

Fig. 4.9 (a) The deep-sea fish *Scopelarchus* showing the upward-pointing tubular eyes, and the wide binocular overlap (B) betwen the monocular fields (M) of the main retinae. L is the lens pad.
(b) Section of the *Scopelarchus* eye, with the outline (dashes) of a conventional eye which has a lens of the same focal length. The main retina images the residual downwelling daylight, whilst the lens pad, composed of light-guiding plates, throws some sort of an image of the dark sector of the field onto the accessory retina. It is a reasonable assumption that the main retina is used to detect silhouettes against the surface, and the accessory retina has the less demanding task of detecting luminescing animals against a dark background. Based on Marshall (1979).

managed to economize on space by using so-called tubular eyes (Fig. 4.9). Optically these are cut-down versions of normal eyes with a reduced, upward-pointing field of view of about 60°, rather than the 180° typical of ordinary fish eyes. By dispensing with the part of the visual field where there is essentially no light, except perhaps the flashes of luminescent animals, the fish manages to incorporate a massive lens into an eye of supportable size. As ever, there are some deep-water cephalopods that have evolved the same trick, for example the tubular-eyed octopus *Amphitretus pelagicus* (Marshall 1979), and the remarkable squid *Histioteuthis* (Fig. 4.5b) which has a large upward-pointing eye, and a smaller downward-pointing one. Deep-sea fish eyes show a range of other curious modifications including multiple banks of receptors, light-emitting photophores in or near the eye, and out-of focus 'accessory' retinas associated with a variety of strange optical devices. In *Scopelarchus* (Fig. 4.9) this structure is called a lens-pad, and behaves as an array of light-guides, but in other species the accessory retina may receive an image of sorts from a mirror, or even from a second lens (Lockett 1977; Collin *et al.* 1998). The probable function of these structures is to provide coverage of the visual field below the animal. This will be dark, with occasional flashes from luminescing animals, and these should be easy to see even with less than perfect optics. In terms of optical reduction the ultimate state is reached by the strange bottom-dwelling fish *Ipnops murrayi*, whose eyes are

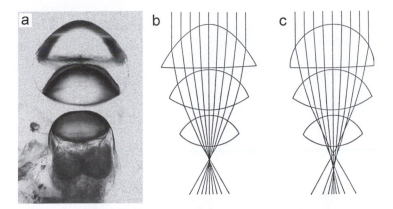

Fig. 4.10 (a) Photograph of the ventral eye of a male *Pontella*, showing the triplet lens and parabolic front surface of the first lens component. For a side view of the animal see Land (1984). (b) Optical construction showing how the parabolic surface enables all parallel rays to come to a point image. Replacing the parabolic surface with a spherical one (c) results in an image that is no longer sharp.

lensless plates of retina, covering the flat upper surface of the front of the head. Undoubtedly these eyes are sensitive, but of what value is sensitivity in the complete absence of resolution?

One particularly interesting recent development is the discovery that at least one deep-water fish (*Aristostomias*) has a red-absorbing visual pigment even though the only light penetrating to those depths is blue, and that it also has a red-emitting photophore. This fish, it seems, has its own private wavelength either for communicating with conspecifics, or lighting up the surroundings as an aid to predation (Partridge and Douglas 1995).

Eyes with non-spherical lenses

There are remarkably few aquatic eyes of the single-chambered type that do not contain single spherical lenses. There are one or two, however, where the required ray-bending is achieved by refraction at a number of surfaces, more in the manner of a multicomponent camera lens. A particularly impressive system is found in the copepod *Pontella*, where a total of six surfaces in three lenses is used to produce the image (Land 1984). This eye is the ventral component of the typical tripartite 'nauplius' eye and, as in other copepods, contains a retina with very few receptors, in this instance only six. This apparent simplicity is the more remarkable because of the amazing development of the eye's optics (Fig. 4.10). In the male there are three lenses, one attached to the eye-cup itself, and another two in the animal's rostrum. In the female, curiously, the most anterior component is missing, making the lens a doublet rather than a triplet. Seen from below it is clear that while most of the surfaces are approximately spherical, this

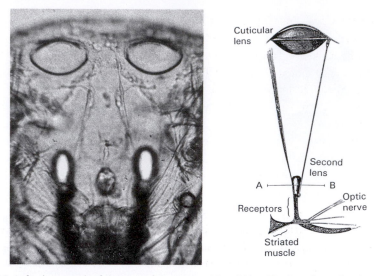

Fig. 4.11 *Left*: photograph of the eyes of the copepod *Sapphirina*. The front lens of each eye throws a large image onto the plane of the second lens which collects light into the small cluster of receptors behind it. *Right*: diagram of the left eye of the related copepod *Copilia*. The striated muscle scans the whole of the rear assembly of receptors and lens back and forth along a track indicated by the line AB. Based on Exner (1891).

is not true of the first surface, which is distinctly parabolic. Ray tracing through the lenses, assuming them to have a uniform refractive index of 1.52, gives the result shown in Fig. 4.10b. With a spherical front surface the system as a whole gives a poor image, with obvious spherical aberration (Fig. 4.10c), but with the parabolic surface this disappears, giving a well-corrected point image. From an optical standpoint, it seems that *Pontella* has hit upon the alternative way of avoiding the perils of spherical refracting surfaces: it has made an aspheric lens, rather than a graded index one.

The eyes of another group of copepods – *Sapphirina, Copilia* and their relatives – have intrigued biologists for well over a century, and it is not hard to see why (Fig. 4.11). Each eye has a pair of lenses. The larger anterior lens is part of the carapace, and it throws an image onto a second smaller lens attached to the front of a tiny retina containing 5–7 receptors. The design is thus somewhat like a pair of telescopes, each with objective and eyepiece lenses. The other reason why these eyes have merited so much attention concerns the way they move. The rear part of each eye, including the second lens, moves sideways in the body, through an angle of about 15°, as measured from the front lens. In *Copilia* the eyes move together, but in opposite directions, at a rate between 0.5 and 10 Hz (Gregory 1991). Although the scanning movements of the eyes increase the effective field of view of the retina, which on its own is only about 3° across, they still only enable the animal to scan a tiny line in the surrounding space. One suggestion has been that *Copilia*'s prey may consist of vertically migrating planktonic animals, which are detected as they swim through the horizontal scan line. This would then provide the point-like retinal detectors with a two-dimensional field of view as in a conventional eye: one dimension resulting from the scanning, and the other from the movements of the prey itself. Unfortunately, there are no direct observations of the behaviour or eye movements of *Copilia* in its natural marine environment.

It is worth mentioning in this context that one fish, the sandlance *Limnichyes fasciatus*, also splits its refraction between four surfaces by making use of a thickened corneal 'lenticle' with a relatively high refractive index (1.38). This tiny but remarkable fish, which has independently-moveable turret-like eyes, catches copepods and other plankton with a rapid, visually-guided lunge. The lenticle can change shape during accommodation and forms part of a very fast focusing mechanism. In conjunction with a rather weak lens, the lenticle brings the nodal point of the optical system towards the front of the eye, thus increasing the focal length and magnifying the image (Pettigrew *et al.* 1999).

Summary

1 Lens eyes evolved from pigmented pit eyes independently in at least 4 phyla.
2 In most cases the lenses have an approximately parabolic gradient of refractive index, falling from a maximum in the centre. This produces a short focal length and a minimum of spherical aberration. Manipulation of this gradient has allowed for partial correction of chromatic aberration in fishes.
3 The large eyes of fishes and cephalopods provide a remarkable instance of convergent evolution. They have separately evolved a variable iris, eye muscles to stabilize the eyes, and focusing mechanisms, but the structure of the retina is totally different in the two groups.
4 The retinae of fish from different environments have specializations in the ganglion cell layer of the retina. Eyes of deep-sea fish often have a tubular shape which permits a large lens in a relatively small eye.
5 In the copepod crustaceans there are a number of examples of compound lenses that use multiple elements and aspheric surfaces, instead of a single inhomogeneous sphere.

5 | Lens eyes on land

A new optical surface

When they emerged from water, the early land vertebrates would have found that their eyes had a new optical arrangement. The cornea, which in water was simply a tough transparent membrane protecting the front surface of the eyeball, became an image-forming structure in its own right, rivalling the lens in its ability to bring rays of light to a focus. In water the cornea has little or no optical effect, because it has a fluid of the same refractive index on both sides. On land, however, the front surface is in air, so there is now a large refractive index difference, across which rays are bent by refraction. It turns out that the ray-bending power of a fish lens and a cornea in air are quite similar. Optical theory states that if the radius of curvature of a surface is r, and the refractive indices on the two sides are n_1 and n_2, then the focal length f of the surface is given by the formula $f = r/(n_2 - n_1)$. This means that there is a focused image of distant objects at a distance f from from the centre of curvature of the surface (see Fig. 5.2 below). For a cornea in air the outside refractive index n_1 is 1, and inside the eye n_2 is about 1.34, so that f becomes $r/0.34$, or about $3r$. In the last chapter we saw that a fish lens of radius r has a focal length of about $2.5r$, so the focal lengths of corneas and lenses with the same radius are quite comparable.

An eye with both a cornea and a fish-type lens has too much focussing power, and if the first proto-amphibian to come on land had done nothing about this it

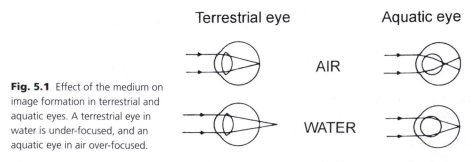

Fig. 5.1 Effect of the medium on image formation in terrestrial and aquatic eyes. A terrestrial eye in water is under-focused, and an aquatic eye in air over-focused.

would have been very myopic (Fig. 5.1). Far away objects would be very blurred, but objects at very much shorter distances, which are focused further from the optics, would be sharp. The blurring would be comparable to what happens to our vision when we go swimming without goggles; in this case, however, we lose the power of the cornea (which now has fluid on both sides) and become hyperopic, which means that we do not have clear vision at any distance.

To look in detail at the ways that animals have adjusted their eyes to life in air, and sometimes to life in both air and water, we will have to deal with combinations of surfaces and lenses. For this reason the next part of this chapter is a primer on the optics of spherical surfaces, which will be useful in trying to understand not only how our single-chambered eyes work but also later when we deal with compound eyes (Chapters 7 and 8). This section ends with a Box in which we work out the focal length and image position in the human eye. This is quite tough going, but for anyone who needs to get to grips with optical systems (biological or otherwise) that make use of multiple surfaces it provides an appropriate tool kit. In the remaining three sections of the chapter we return to biology, and examine first the range of vertebrate eyes that use corneal optics, exploring such topics as focussing mechanisms and ecological adaptations in the process. There is then a short section on amphibious eyes that have to be made to function in both air and water. Finally we explore the eyes of those invertebrates, principally the spiders, that also employ a cornea to form their images.

Basic optics of cornea and lens

To work out how an eye will perform we usually want to know where the image is and how large it is, for an object whose size and distance are known. For a single curved surface there are well-known formulae for making these calculations, and these are given here. Where more surfaces are involved, the calculations can become very complicated. However, it is usually possible to

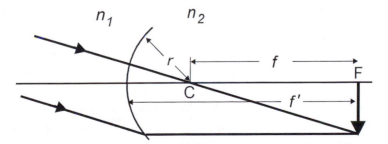

Fig. 5.2 Image formation by a curved cornea. An image of a distant object is formed at F, situated a distance f from the centre of curvature C, and f' from the surface itself. These distances are given by eqns (5.1) and (5.2); r is the radius of curvature of the surface, which separates media with refractive indices n_1 and n_2.

'reduce' a complex image-forming structure to a single equivalent surface with the same optical power, and then the simple formulae can be applied. In Box 5.1 we show how this can be done for the human eye, using methods that can be applied to any eye with a combination of curved surfaces and lenses.

Curved surfaces separating two media of different refractive index form images, provided the higher refractive index is on the concave side (Fig. 5.2). The position of the image of an object at infinity can be calculated either from the centre of curvature of the surface, or the surface itself. If we consider an eye whose main optical surface is a cornea separating two media of refractive indices n_1 and n_2, these two distances, f and f', are given by:

$$f = n_1 r/(n_2 - n_1) \tag{5.1}$$

$$f' = n_2 r/(n_2 - n_1) \tag{5.2}$$

where the radius of curvature of the surface is r. Notice that $f'/f = n_2/n_1$. When the first medium is air ($n = 1$), which is the usual case, these equations become:

$$f = r/(n_2 - 1) \tag{5.1a}$$

$$f' = n_2 r/(n_2 - 1). \tag{5.2a}$$

Although it seems more sensible to measure image position from the surface itself (f'), it turns out that f (also known as the first focal length or the posterior nodal distance) is the more useful measure. The reason for this is that rays of light passing through the centre of curvature are not bent by the surface, because they cross it at right angles (Figs 5.2 and 5.3). This means that an object in the outside world, and its image behind the surface, make the same angle at the centre of curvature, and this in turn means that one can use the principle of similar triangles to work out the size of the image of any object. In Fig. 5.3, if object and image sizes are O and I, and their distances from the centre of curvature are U and V, then:

$$I/O = V/U. \tag{5.3}$$

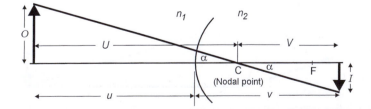

Fig. 5.3 Relations between object (O) and image (I) at a curved surface. A ray passing through the centre of curvature (C) is not bent by the surface, so image and object are related by the two similar triangles with angle α. Sizes and distances of the object and image are related by eqns (5.4) and (5.6).

The ratio of the sizes of object and image (I/O) is often referred to as the magnification, m. If the object distance U is large (say >100 times V) then the image will be very close to the focal point for an object at infinity, so V can be replaced by f, in which case:

$$I/O = f/U. \tag{5.4}$$

O and U are usually easily measured, so that the image size I can be found if f is known. This equation applies to any optical system, provided the right value for f is used.

For example: a grating of black and white lines, with a spacing between the black lines of 5 mm, is just distinguishable from grey at 17 m. How far apart are the images of the lines on the retina? We know that the focal length of the average human eye (the cornea/lens combination) is 16.8 mm. Applying eqn (5.4) then gives $I = 5 \times 16.8/17\,000$ mm, or 4.9 μm. This is approximately twice the separation of cones in the fovea, so that each black/white stripe pair in the image has two cones to receive it.

An important concept in dealing with any optical system is the *nodal point*. This is a point on the axis and is defined as the point of intersection of straight lines connecting points on the image with points on the object (which is assumed to be at a large distance). These may be rays in simple systems, but in more complex systems this is not necessarily the case (Fig 5.4); however, the definition just given still applies. The distance from the nodal point to the image of a point at infinity is by definition the focal length f (Fig. 5.4). Sometimes it is easy to decide where the nodal point is. In a simple refracting cornea it is at the centre of curvature, as we have seen. For a thin lens in air the nodal point is at the centre of the lens. For a fish lens in water it is again at the lens centre, because the lens is spherical. However, for more complex systems involving thick lenses it is necessary to find the posterior nodal point by ray-tracing, i.e. working out where rays

Fig. 5.4 Definition of focal length *f*. Rays from a distant object making an angle α with the optical system produce an image of size *I* at F. The ray in image space that makes an angle α with the axis and passes through the image, must also pass through the nodal point N. The distance from N to F is the focal length *f*. If α is small then α (in radians) = *I/f*. If α is known from measurements in object space, then *f* can be found from the image size *I*.

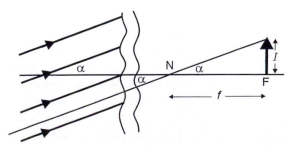

go surface by surface, and a method for doing this is given later in this section. In the human eye the situation is fairly simple: for most purposes the system can be regarded as having a nodal point 16.8 mm in front of the focus for distant objects.

Besides making it possible to use eqns (5.3) or (5.4) to work out image sizes from object sizes, the nodal point also allows one to specify both image and object sizes in terms of the angles (α in Figs 5.3 and 5.4) that they subtend at the nodal point, as these angles are the same. If O is small compared with U, then O/U is an angle in radians (if $O = 1$ and $U = 10$ then the angle is 0.1 radians, which is $0.1 \times 180°/\pi$, or 5.73°).

For objects that are closer to the eye the image moves deeper, behind the focal point for rays from infinity (for optical purposes infinity is a few metres away for an eye like ours) . Equation (5.4) does not apply, and f must be replaced by the actual image distance. This is calculated from:

$$n_2/v - n_1/u = (n_2 - n_1)/r = 1/f, \tag{5.5}$$

where the object and image distances u and v are now measured from the surface itself, not the nodal point (this is why the symbols have been changed from upper to lower case; $u = U - r$, $v = V + r$ in Fig. 5.3). With air outside, eqn (5.5) becomes:

$$n_2/v - 1/u = (n_2 - 1)/r = 1/f. \tag{5.5a}$$

It is important in using this equation to stick to an appropriate 'sign convention'. The most straightforward is the Cartesian convention familiar from graphs, where distances to the right of the surface are taken to be positive. So in Fig. 5.3 u is negative and v, r, and f are positive. If the refracting surface were concave to the left, however, r would become negative, as would f. To work out the image size using u and v, the corresponding equation to (5.3) is:

$$I/O = (v/n_2)/(u/n_1), \tag{5.6}$$

or if the object is in air,

$$I/O = v/un_2. \tag{5.6a}$$

As an example, if a book is at 40 cm from a human eye, which makes no focusing effort (perhaps because the reader is over 50), how far behind the eye's focal plane will the text be focused? We require v from eqn (5.5a), where u is –400 mm and f has a typical value of 16.8 mm (we are assuming here that the eye consists just of a cornea with this focal length, and an internal refractive index of 1.336). Rearranging eqn (5.5a) gives $v = n_2/(1/f + 1/u)$, from which $v = 23.43$ mm. The focal point for an object at infinity is $f' = n_2f$ behind the front surface, i.e. 22.51 mm, so if the retina is in focus for objects at infinity, the image of the text will lie 0.92 mm behind this. This will produce a seriously degraded image. With a 3 mm pupil, the image of a point source on the retina will be a blur circle about 0.12 mm wide – half the width of the fovea.

In dealing with systems that have more than one surface, it is often convenient to work with *powers* rather than focal lengths. These are reciprocals of focal lengths ($1/f$) and they can be added together. Thus if two adjacent surfaces, or thin lenses, have powers P_1 and P_2 (i.e. $1/f_1$ and $1/f_2$), the combined power P_{comb} is given by:

$$P_{comb} = P_1 + P_2 \tag{5.7}$$

and the corresponding focal length is $1/P_{comb}$. The unit of optical power is the dioptre (symbol D), which is $1/f$ when f is measured in metres. The optics of a typical human eye give a focal length of 16.8 mm, corresponding to a power of nearly 60 D. Of this the corneal power is about 40 D and the lens 20 D. In optometry powers can be used to specify both optical defects and the lenses needed to correct them. Thus if someone has 5 D of myopia, meaning that the optics focus too far forward, this will require a –5 D negative lens to correct it.

When the elements of an optical system are separated, as they often are in real eyes, the power of the system is somewhat less than simple addition suggests. For separated surfaces the power formula becomes:

$$P_{comb} = P_1 + P_2 - dP_1P_2/n \tag{5.8}$$

where d is the distance between the surfaces, and n is the refractive index in that space. The powers of the individual surfaces can be found from their focal lengths, by applying eqn (5.1.)

For systems that are more complicated than a single thick lens, to which eqn (5.8) applies, the position of the focus can be found by ray tracing. Basically, one starts with parallel rays coming from the left (as in Fig. 5.2) and applies eqn (5.5) to each surface in turn. When the image formed by one surface is located, it becomes the object for the next, and so on. This is foolproof, but sticking to the sign convention is crucial for success! In Box 5.1 this method is applied to the human eye. It can be used to determine the imaging properties of any eye of the lens/cornea type.

Box 5.1 A model of the human eye

We can illustrate how to use ray tracing to find the position of the image in the human eye, and its focal length, with a model that has served optometrists and ophthalmologists well for nearly a century. It rejoices in the name of the Gullstrand simplified (No. 2) schematic eye, named after the Swedish ophthalmologist who devised it, and is illustrated in Fig. 5.5a. Similar model eyes have been devised for a number of vertebrates, includ-

Box 5.1 A model of the human eye (*contd.*)

ing the goldfish, frog, turtle, pigeon, rat, cat, and monkey (Charman 1991; Martin 1983). The Gullstrand model consists of a cornea and a lens, all having radii of curvature close to those of real eyes (see Fig. 5.5a). There are thus three surfaces to consider. (The human lens, like fish lenses, is not homogeneous, but Gullstrand chose a single refractive index of 1.413 that would provide a homogeneous lens with the same power as a real lens.) To determine the image position relative to the rear surface of the lens we apply eqn (5.5) to each surface in turn. In this calculation the subscripts of u, v, and r all refer to the surface (1–3) at which refraction takes place. Refractive indices are labelled from air on the left (1) to the vitreous space behind the lens (4). For light rays from infinity reaching the cornea we have:

$$n_2/v_1 - n_1/u_1 = (n_2 - n_1)/r_1$$

i.e.

$$1.336/v_1 - 1/\infty = (1.336 - 1)/7.8,$$

hence

$$v_1 = 31.01 \text{ mm.}$$

The object distance for the next interface (u_2) is shorter than this by the distance separating the cornea and the front of the lens (3.6 mm) so that $u_2 = 27.41$ mm. So at the front surface of the lens:

$$n_3/v_2 - n_2/u_2 = (n_3 - n_2)/r_2$$

i.e.

$$1.413/v_2 - 1.336/27.41 = (1.413 - 1.336)/10.0,$$

hence

$$v_2 = 25.03 \text{ mm}$$

The front and rear lens surfaces are also separated by 3.6 mm, so the object distance u_3 for the next surface is 21.43 mm. At the rear lens surface:

$$n_4/v_3 - n_3/u_3 = (n_4 - n_3)/r_3$$

i.e.

$$1.336/v_3 - 1.413/21.43 = (1.336 - 1.413)/-6.0$$

hence

$$v_3 = \mathbf{16.96} \text{ mm,}$$

Box 5.1 A model of the human eye (*contd.*)

which is the distance from the rear surface of the lens to the focus.

In a system of several surfaces, the distance from the rear surface to the focus is not the focal length (f or f'), although in this case it is fortuitously similar to f. The focal length (f) is defined by the magnification of the image. If an object at infinity subtends an angle α at the eye, then the formula $I/f = \tan \alpha$ (which is essentially the same as eqn 5.4) gives the *equivalent* focal length, for any image-forming system (Fig. 5.4). The method is to work out the size of the final image by taking the size of the initial image and multiplying it by the magnifications m of each succeeding surface using eqn 5.6

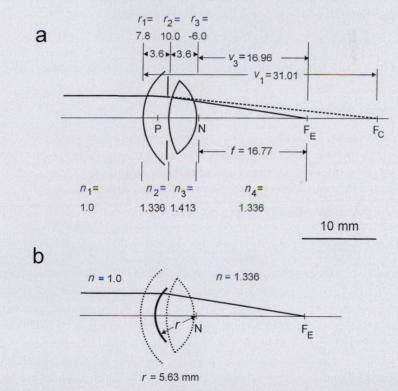

Fig. 5.5 (a) Dimensions of the Gullsrand (No. 2) schematic human eye. P, principal plane; N, nodal point: F_E, focal point of the eye, for an object at infinity; F_C, focal point of the cornea on its own. Further explanation in the text. (b) Reduced eye. For most purposes the optical system in (a) can be replaced by a single refracting surface, radius r, centred on the nodal point. The surface is situated at the principal plane in (a).

Box 5.1 A model of the human eye (*contd.*)

$(m_k = I/O = (v_k/n_{k+1})/(u_k/n_k)$, at the kth surface). All the values for the lengths involved are available from the preceding calculation of the position of focus. From the definition of focal length just given:

$$I_f/f_e = \tan \alpha$$

where I_f is the size of the image formed by the final (3rd) surface, and f_e is the equivalent focal length of the whole system. Working through the interfaces, beginning with the cornea we have:

$$I_1/f_1 = \tan \alpha$$

where $f_1 = f_1'/n_2 = v_1/n_2$, so that $I_1 = v_1 \tan \alpha/n_2$. At the next interface $I_2 = m_2 I_1$, where $m_2 = I_2/O_2 = (v_2/n_3)/(u_2/n_2)$. Similarly at the third and final interface: $I_3 = m_3 I_2$, where $m_3 = (v_3/n_4)/(u_3/n_3)$. The final result is that

$$I_3 = v_1 \tan \alpha/n_2 \,.\, (v_2/n_3)/(u_2/n_2) \,.\, (v_3/n_4)/(u_3/n_3).$$

Then, since $f_e = I_3/\tan \alpha$, the tan α terms cancel, and the final expression for f_e is:

$$f_e = v_1/n_2 \,.\, (v_2/n_3)/(u_2/n_2) \,.\, (v_3/n_4)/(u_3/n_3)$$

which reduces to

$$f_e = (v_1 v_2 v_3) / (u_2 u_3 n_4).$$

Substituting the values from the previous calculation gives:

$$f_e = (31.01 \,.\, 25.03 \,.\, 16.96) / (27.41 \,.\, 21.43 \,.\, 1.336)$$

$$f_e = \textbf{16.77 mm}.$$

We now have the position of the image, 16.96 mm behind the rear surface of the lens, and also the equivalent focal length, 16.77 mm. Notice that these calculations show that the nodal point of the eye is just behind the rear surface of the lens. For most practical purposes, this is all one needs to know about an optical system to work out where images will fall, and how big they will be. The focal length can be used to work out the sizes of images using eqn (5.4), and changes in image position with object distance can be found from eqn (5.5), taking n_2 as 1.336, the refractive index of the rear chamber of the eye.

What we have effectively done is to reduce the rather complex optics of the human eye to a single air–fluid interface with a focal length of 16.77 mm. Because the refractive index is specified (1.336) the radius of curvature of

Box 5.1 A model of the human eye (*contd.*)

the fictitious surface is given by eqn (5.1a), and it comes to 5.63 mm, rather less than the actual cornea, because, of course, it has had the power of the lens added to it. The surface must be situated a distance r in front of the nodal point, i.e. 22.40 mm from the image, and its position on the axis is known as the principal point of the system. This 'reduced' eye is shown in Fig. 5.5b.

The methods given above can be applied to any eye that uses lenses or lens–cornea combinations, including the ommatidia of apposition compound eyes.

Variations on the lens/cornea theme in land vertebrates

To solve the problem of excessive optical power, land vertebrates could have done a number of things. They might have abandoned the lens altogether and adopted the cornea as the sole image-forming structure, or they could have kept the lens and flattened the cornea so that it had no power, or they could have retained both but shrunk the eye to fit the shorter focal length of the combined system. The last of these possibilities, or something like it, does occur in nocturnal mammals; but most of the reptiles, birds and mammals have opted for a compromise, in which the lens is retained, but with much less power than the ancestral fish lens. The lens and cornea then divide the optical power between them: in humans this ratio is about 1 to 2. In optical technology it is usually a good idea to split the required refraction between several surfaces, because the optical defects (aberrations) of several weakly curved surfaces are usually less than those of a single surface of much stronger curvature. Retaining a weaker, flatter lens, together with a not too curved cornea, may have been a way of obtaining images of high quality.

Sizes and shape of eyes

Figures 5.6 and 5.7 illustrate the variety of eye shapes among land vertebrates. The basic design of the eyes of reptiles, birds and mammals is very similar, with a hemispherical rear chamber, a biconvex lens, and a cornea with a pronounced curvature. There are differences in the 'housekeeping arrangements' between the classes; for example, the nature of the vascular supply varies and so does the mechanism of accommodation. However, the main differences that are apparent in Fig. 5.7 are not linked to phylogeny but to lifestyle. These major variations are of three kinds: there are differences in overall size, differences in the shape and relative size

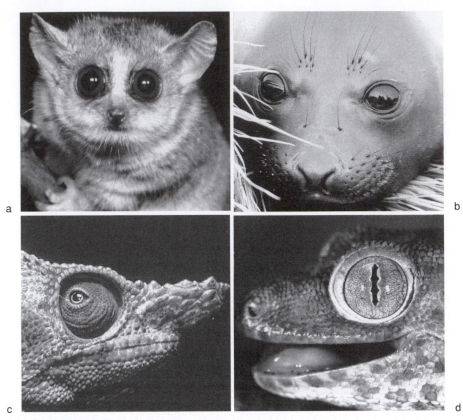

a
b
c
d

Fig. 5.6 Photographs of the eyes of various terrestrial vertebrates. (a) Mouse lemur (*Microcebus murinus*): typical large nocturnal eyes with wide pupils. (b) Elephant seal (*Mirounga leonina*): flattened corneas are an adaptation to amphibious life (see Fig. 5.8). (c) Chamaeleon (*Chamaeleo oshaughnessyi*): turret-like diurnal eye with a small pupil. (d) Tokay gecko (*Gecko gecko*): active day and night; notched slit pupil partially open (see Fig. 5.11). Photographs: (a) and (c) David Haring, Duke University Primate Center; (b) Johnny Johnson (Bruce Coleman Collection); (d) Kim Taylor (Bruce Coleman Collection).

of the lens related to nocturnal life style, and a tendency to redevelop a spherical lens and a flat cornea in species that have returned to an aquatic life.

Why are some eyes bigger than others? The answer seems obvious: bigger animals have bigger eyes. This is true to some extent (see Hughes 1977, p. 654) but it is certainly not the whole answer. Birds tend to have much bigger eyes for their body size than mammals, and smaller mammals have relatively bigger eyes than larger ones. Zebras, for example, have larger eyes than elephants, and even some whales. The largest eye of any land animal is that of the ostrich, with a diameter of 50 mm compared with 40 mm for a horse and 24 mm in man (Martin 1985).

There are two optical reasons for having a large eye: resolution for acute vision, and sensitivity for vision in dim conditions. Good resolution requires a

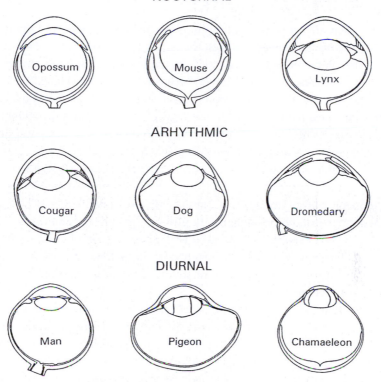

Fig. 5.7 Variations in the structure of eyes from animals with different terrestrial lifestyles (not on the same scale). Nocturnal animals have the biggest lenses and diurnal animals the smallest; animals active day and night (arhythmic) have intermediate eyes. Adapted from Walls (1941).

large eye to provide a long focal length, so that the angle between receptors is as small as possible (see Chapter 3, eqns 3.1 and 3.2). This has to be matched by good image quality, which requires a large lens to provide a small diffraction blur-circle (see Chapter 3, eqn 3.3). Both these conditions imply that the larger the eye the better the resolution, and this is why primates and birds of prey have large eyes. Horses, however, have large eyes, but do not have the same need for high resolution vision, and indeed their resolution is not remarkable, about 2.5 times worse than man. The other explanation must be that horses are partly nocturnal. The large eye is then needed to achieve a wide aperture for capturing photons, as discussed in Chapter 3. People familiar with horses say that they can pick their way through difficult terrain at light levels where the rider can barely see the ground. Animals that need both high resolution and high sensitivity have particularly large eyes. The reason for the 'tubular' shape of owl eyes is that hemispherical eyes with such long focal lengths and wide apertures would not fit into the head. Owls have squeezed them in by removing much of the peripheral part of the globe, but there has been a price to pay; owl eyes cannot move more than a few degrees around any axis, despite a full set of eye muscles.

Animals that are active day and night, such as horses, owls, and many mammalian carnivores, have eyes in which the lenses have a diameter of about 0.4 to 0.5 times the diameter of the eye itself. In truly diurnal animals, for example monkeys and parrots, the ratio is lower, between 0.3 and 0.4 (Fig. 5.7). However, in nocturnal animals that rarely emerge in daylight, such as the house mouse, opossum, and bush baby the lens diameters are 0.6 to 0.8 times the eye diameter (Fig. 5.6a). These differences are of relative not absolute lens size, and are concerned with getting as bright an image as possible for a given size of eye. A large almost spherical lens, combined with a strongly curved cornea, gives a very short focal length, and combined with a wide aperture this gives the eye a very high light gathering power (image brightness is proportional to $(D/f)^2$, where D is aperture diameter and f focal length). In photographic terms a house mouse has an *F-number* (f/D) of about 0.9, compared with about 2.0 for a human with a wide open pupil. The mouse's image is brighter by a factor of nearly 5. Generally speaking the power of the cornea is relatively more important in diurnal eyes, and the lens in nocturnal eyes.

Animals such as seals have spherical lenses for a different reason. Their problem is that, having returned to water, they no longer have an optically useful cornea. They therefore require a much more powerful lens, and that means a spherical lens, just as in fish (Fig. 5.8). However, they are not wholly aquatic, and when they come onto land the re-appearance of a strong cornea would make them very myopic. One solution is a flattened cornea with little power in either medium, and as the comparison with the lynx shows, this is the direction that the seal eye has gone (Fig. 5.6b). Incidentally this is one reason why baby seals look so appealing. Their eyes are like limpid pools, not because of the purity of their souls, but because the corneas are flat.

Fig. 5.8 Return to the sea. The eyes of seals have a flattened cornea and a spherical lens, like the eyes of fish (see Fig. 4.6). Their terrestrial relatives, such as the lynx, have a domed cornea and a thinner, weaker lens. From the famous set of engravings of the eyes of mammals, birds and fish by D.W. Soemmerring (1818) *De Oculorum Hominis Animaliumque*. Vandenhoek & Ruprecht, Göttingen. The drawings are all of the lower hemisphere of the left eye.

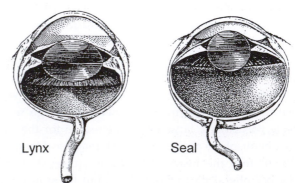

Lynx   Seal

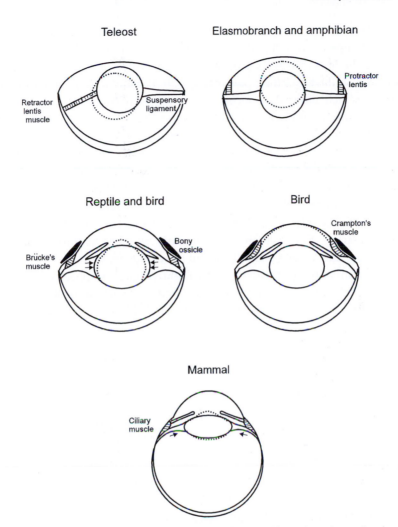

Fig. 5.9 Accommodation mechanisms in vertebrates. The dashed lines show the results of contraction of the accommodatory muscles. In teleosts the lens is drawn bodily towards the retina, accommodating for distant objects. In elasmobranchs and amphibians the lens moves away from the retina, accommodating for near objects. In reptiles and birds the lens can be deformed by Brücke's muscle which pushes on the lens via the ciliary body. In birds contraction of Crampton's muscle pulls on the cornea, decreasing its radius of curvature. In mammals the lens also deforms, but this occurs by the elasticity of the capsule around it. Contraction of the ciliary muscle relaxes the tension on the structures supporting the lens, allowing it to deform. In reptiles, birds, and mammals the actions of the accommodation muscles all permit near objects to be brought into focus.

Accommodation, a new function for the lens

In fish, with a spherical lens of fixed focal length, the only available mechanism for focussing is for the lens to move bodily towards or away from the retina, just as in a camera. In mammals, birds and reptiles, however, the lens is deformable and so can change its focal length. In mammals, relaxation of the elastic capsule surrounding the lens causes its surfaces to bulge, which decreases their radii of curvature and so increases their power (Fig. 5.9). Paradoxically, the relaxation of the capsule – which allows close focusing – comes about by an increase in tension in the ciliary muscle, and hence the eye is focused for distance when the muscle is relaxed. In a young person the radius of curvature of the front face of the lens can halve, from 10 mm to 5 mm, although the change in the rear surface is much less pronounced. The result is an increase in the power of the eye as a whole by 8.6 dioptres. This is equivalent to putting a lens with this power in front of the unaccommodated eye (eqn 5.7), and as such a lens would have a focal length $(1/P)$ of 0.116 m, the effect is to enable the eye to focus on a point 11.6 cm away. Sadly, lens elasticity declines more or less linearly with age, and for most people the lens has lost all focusing power by about the age of 55.

Accommodation in reptiles and birds is slightly more complicated. One set of muscles (Brücke's muscle) pushes the ciliary body inward on the lens as it contracts, so deforming it and increasing its power, while another muscle system (Crampton's muscle) pulls on the cornea in a way that reduces its radius of curvature. The corneal mechanism is particularly important in birds. Chamaeleons, which use their focusing mechanism to judge the distances of insect prey, accommodate particularly fast, with a mechanism based on lens deformation (Ott and Schaeffel, 1996). Remarkably, the chamaeleon lens at rest has *negative* power, implying a refractive index profile quite different from the usual 'highest in the centre' gradient. This gives the combined optical system of cornea and lens a very long focal length, and a correspondingly magnified image. Other features related to the chamaeleon's lifestyle, which involves searching for insects amongst foliage, are the extreme mobility of the turret-like eye (Fig. 5.6c) and the presence of a distinct high-resolution fovea.

A type of accommodation that involves no movements or deformations of the lens is apparently found in horses and rays (elasmobranch fishes). The mechanism is known as a 'ramp retina' and works on the principle that the upper part of the retina will always be imaging the lower (closer) part of the field of view, whereas the lower retina will image distant objects. The upper retina thus needs to be further away from the nodal point than the lower part, leading to a retina to whose spherical shape has been added a backward slope, or ramp. Doubts have been raised about the reality of ramp retinas, because the differences in retinal distance involved are much smaller than the early diagrams suggested (Sivak 1976). Nevertheless, 'lower field myopia' is well established in ground foraging birds, where it keeps the retina focussed on the ground plane, and birds appear to have mechanisms that do indeed adjust the relative positions of retina and focal plane during development (Schaeffel *et al.* 1988). Another 'static' form of accommodation is found in fruit-bats. Here the retina is deformed by a series of conical papillae that ensure that adjacent local regions are at different distances from the lens, and so are in focus for objects at different distances.

Corneal shape and spherical correction

Spherical aberration is not just a problem for lenses, as we saw in Chapter 4. It afflicts spherical air–water surfaces in much the same way, overfocusing rays more and more as the distance from the axis increases. One cure for corneal spherical aberration is to make the surface aspherical, with the outer regions having lower curvature (and hence relatively less power) than those close to the axis. This is what happens in the human eye; the periphery of the cornea has a radius of curvature twice that of the central region. The overall shape that produces a point image has an elliptical profile, and the human cornea approximates to this.

There is a price to be paid for having a non-spherical cornea. Because the spherical symmetry of the old aquatic eye has been lost, the aspheric eye has a single optical axis along which the optics are corrected, and image formation is particularly good. However away from this axis the cornea presents a tilted profile and the image quality gets rapidly worse. A consequence of this is the highly centred visual system of primates, including man, where 'good' vision is concentrated in a central foveal region only 1° across. To use this effectively we have a very sophisticated eye movement system that finds objects of interest in the periphery and centres them for foveal scrutiny (see Chapter 9). Fixation of this kind is relatively uncommon, even amongst mammals.

It seems that the lens in man corrects its own aberration in the same way as the lenses of fish (Chapter 4), by having a refractive index gradient (Millodot and Sivak 1979). It is, after all, their descendant. So each refracting structure looks after itself: the cornea by being aspheric and the lens by being inhomogeneous. Something rather more interesting occurs in the rat, and other mammals with nocturnal-type eyes in which the large spherical lens forces the cornea to be more or less spherical too (Chaudhuri et al. 1983). The cornea has thus little scope for an aspheric correction – which might in any case be inappropriate because most of these animals need a large field of view without major variations in image quality across it. Instead the lens has a gradient that makes it over-corrected, like a fish lens but more so, so that it corrects not only its own aberration but that of the cornea as well. And because the system is spherical the correction works across the whole field.

Form and function of the pupil

Most land vertebrates have a very active pupil that changes size rapidly in response to changes in illumination of the retina. This is in contrast to the situation in many fish, where the pupil, if active at all, may take minutes to open and close, and is often activated directly by light rather than neurally, via the retina.

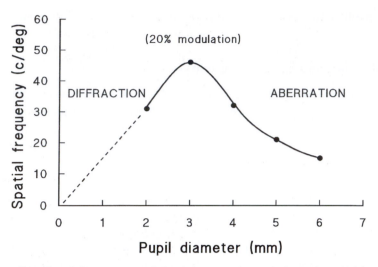

Fig. 5.10 Effect of pupil diameter on resolution in humans. The graph shows the spatial frequencies that provide 20 per cent modulated images at different pupil diameters in the human eye, which is a convenient measure of resolution. The optimum pupil diameter is 3 mm. Data from Charman (1991).

The human pupil changes in diameter from about 2 mm in sunlight to 8 mm in the dark, giving a maximum theoretical difference in sensitivity of 16 times, though for various reasons this reduces in practical terms to about 10 times. Compared with the full range of lighting conditions over which the eye operates (about 10^{10}) this is very little. It seems that the function of the pupil in man is not so much to compensate for changes in brightness as to obtain the best compromise between resolution and sensitivity. In bright light the pupil diameter of 2–3 mm is optimal in the sense that it provides the best resolution the optics can support (Fig. 5.10). Smaller apertures result in a poor perfomance because diffraction reduces the spatial cut-off frequency (Chapter 3), and with apertures bigger than about 3 mm other defects such as spherical aberration become progressively worse, again reducing resolution. As light levels drop, resolution starts to become limited by photon noise rather than optical quality (Fig. 3.10), and it then pays to open the pupil to admit more light, as the resulting decrease in optical acuity will no longer be noticed.

Nocturnal animals have wide aperture eyes that are intrinsically more sensitive than our own, and these require protection in daylight to prevent the bleaching of all the photopigment. The muscular mechanics of the circular pupil mean that it cannot close down beyond a certain limit, and the alternative is a slit pupil, which can close much further. In the cat eye for example (Fig. 5.11) the change in pupil area between dark and light is 135-fold, a ten times greater range than in man. In gecko eyes the two margins of the slit have a series of paired notches (Fig. 5.6d), and when the pupil closes these match up to give a

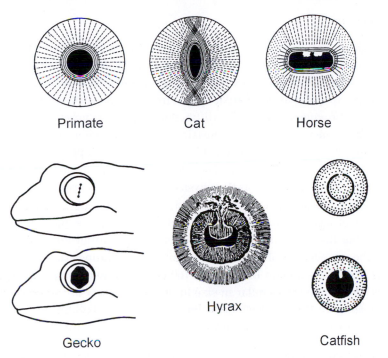

Fig. 5.11 Pupil shapes in vertebrates. *Top row*: round and slit-shaped pupils in mammals, showing how the cat's slit pupil can close further than the circular primate pupil. Iris closer muscles are continuous lines and opener muscles dashed lines. *Bottom row*: gecko pupil contracts to four 'pin-holes' in the light. The hyrax or coney (*Procavia*, a small desert mammal) has a pupil partly closed by a central operculum, which acts as a sunshade. A similar mobile operculum is present in some fish, such as the catfish *Plecostomus*. Combined from Walls (1941).

set of small pin-holes (Fig. 5.11). Compared with the fully open nocturnal pupil this cuts down the light by more than a thousand-fold, enabling geckos to hunt in daylight without damage to the retina. Slit pupils may be horizontal, as in some sharks, or more commonly vertical, as in many lizards, snakes, and mammals. Horizontal pupils in mammals tend to be broadly oval, as in horses (Fig. 5.11) and ruminants (Walls 1942). *Octopus* also has an oval horizontal pupil (Plate 1).

Another pupil arrangement that permits a high degree of closure is a circular ring with an expandable operculum inside it. This is quite common in shallow-water fishes (e.g. rays and catfish, Fig. 5.11, Plate 1), and amongst mammals in the hyrax (*Procavia*) and some whales. This arrangement has the additional advantage of acting as a sunshade, excluding strong light from above, and in some cases it may also act to camouflage the eye. Squid and cuttlefish have W-shaped pupils, possibly for the same reasons. Calculations suggest that they also provide a more even retinal illumination than a circular pupil.

Resolution

The two factors that limit an eye's resolution are the quality of the optics, and the fineness of the retinal mosaic, as was discussed in detail in Chapter 3. Psychophysical measurements show that the finest detail a well-focused human eye can resolve, expressed as an angular spatial frequency, is very similar to both the optical cut-off frequency set by the diffraction limit (D/λ cycles/radian) and the sampling frequency of the retinal mosaic ($f/2s$ c/rad), where D is the aperture diameter, λ the wavelength, f the focal length and s the receptor separation. Both are close to 60 c/deg (3438 c/rad). This means that the performance of a human eye, in bright light, is as good as the physical constraints on optics will allow, and that the retinal mosaic has evolved to match this optical limit.

In terms of resolution, our eyes are probably at least as good as any other mammal, but they are not quite as good as some raptorial birds. Behavioural data for the some hawks shows that they can resolve gratings that are more then twice as fine as the human limit, about 160 c/deg. How is this achieved, given that the hawks' eyes are similar in size to our own? There are three differences that contribute to this improvement. First, the hawk's daylight pupil is wider than ours, about 6 mm, which improves the diffraction limit by a factor of betwen 2 and 3. Second, the foveal receptors are narrower, about 2 μm between centres, rather than 2.5 to 3 μm in man. Thirdly, the hawk uses an optical trick to increase the effective focal length of the eye. It seems that the pit-like depression in front of the fovea acts as a negative lens, since the material of the retina has a higher refractive index than the vitreous humour, thereby creating a modest telephoto system, the principle of which is shown in Fig. 5.12. The image focused by the cornea and main lens is shifted backwards by the negative lens, giving a longer overall focal length, and a locally magnified image. The magnification of the system, relative to a system without the negative lens, is about 1.45 (Snyder and Miller 1978). The overall effect of these various modifications is a linear resolution gain of more than 2, but in terms of the number of foveal receptors imaging a given area, it is a gain of 4.

Other vertebrates may or may not have as good a match between the diffraction and sampling limits as do hawks and humans. Many are nocturnal, and use their wide pupils for sensitivity, not resolution. The hooded rat, for example, has an aperture of about 0.3 mm in bright light, roughly 10 times smaller than man (Hughes 1977). This implies a diffraction limit to resolution of about 10 c/deg. In fact the threshold measured by behavioural testing is about 1 c/deg, 60 times worse than man. This is a much larger discrepancy than can be accounted for by optics alone. The horse, with an eye nearly double the diameter of ours has a behavioural resolution (measured by its ability to distinguish stripe patterns from uniform grey) of only 23 c/deg, about three times worse than man (Timney and Kiel 1992). The lower resolution, relative to the diffraction limit, of these and

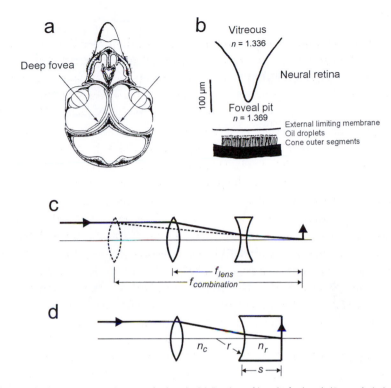

a Deep fovea

b Vitreous
$n = 1.336$

Neural retina

100 µm

Foveal pit
$n = 1.369$

External limiting membrane
Oil droplets
Cone outer segments

c

f_{lens}
$f_{combination}$

d

n_c r n_r

s

Fig. 5.12 Telephoto optics in the eye of a hawk. (a) Section of head of a hawk (*Buteo latissimus*) showing the eyes meeting on the centreline (typical of birds) and the direction of view of the deep foveas. Temporal foveas are also shown. (b) Diagram of the foveal pit and its relation to the retina. In the telephoto theory the optically important surface is the spherical bottom of the foveal pit. (c) Construction of a telephoto camera lens. The effect of the negative (concave) lens is to increase the focal length of the combination, so that its imaging properties are those of a single lens in front of the combination (*dotted*). (d) As above, but with a single concave surface, corresponding to the foveal pit. The magnification of the system is given by: $m = 1 + (s/r).(n_r - n_c)/n_c$. Redrawn from Snyder and Miller (1978).

many other vertebrates may be due to optical imperfections, larger receptors, or commonly the grouping of receptors into larger units based on ganglion cells, within which there is no further resolution.

Ecology, resolution, and ganglion cell distribution

Although this chapter is not specifically concerned with the organization of the retina, a consideration of the way ganglion cells are distributed in vertebrate eyes tells us a good deal about an animal's visual priorities. Animals that live a life dominated by activity around the horizon, for example predators like the cheetah and herbivores such as rabbits and ungulates, have a narrow horizontal strip

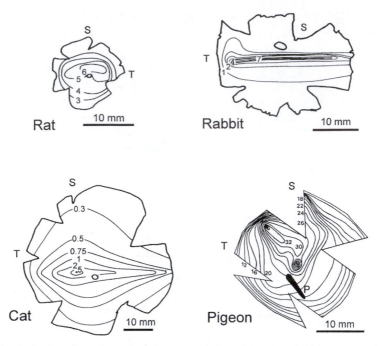

Fig. 5.13 Distribution of ganglion cells in the retina. The ganglion cells supply the axons in the optic nerve, and so represent a 'bottleneck' in the visual system. Numbers are thousands of nuclei per square millimetre of the flattened retina. The rat has a roughly circular area centralis, whereas the rabbit has a linear visual streak corresponding to the horizon. The cat is somewhat intermediate. The pigeon has two distinct foveas with particularly high ganglion cell densities. One looks laterally along the optical axis of the eye, whereas the other is situated temporally, and images the region of the bill tip. T and S, temporal and superior. P, the position of the pecten, a nutritive structure in the bird eye. Modified from Hughes (1977) and Martin (1985).

through the retina where the ganglion cell density is very high – the 'visual streak' (Fig. 5.13). Animals from a more three-dimensional environment, such as forest, either have a uniform retina, or one with a more or less circular 'area centralis' where ganglion cells are concentrated (Hughes 1977). A similar situation occurs in fishes from different underwater niches, as we saw in Chapter 4 (Fig. 4.7). In primates this 'area' concentrates to a 1° central spot, the fovea centralis, with exceptionally high numbers of ganglion cells associated with it – one per cone, about 150.10^3 per square millimetre. This compares with about 6.10^3 in the centre of the area centralis of the rat retina. Birds often have two foveas, one looking out laterally, and the other, situated at the rear (temporal region) of the retina, imaging the region of the bill where the bird pecks at food (as in the pigeon retina in Fig. 5.13).

This tendency for ganglion cell densities to match the pattern of 'interest' in the environment is seen in mammals, birds (Martin 1985) and fish (Collin and

Pettigrew 1988). It seems to be a way of economizing on the numbers of axons that have to leave the eye for the brain, by matching visual information to neural capacity. In vertebrates the optic nerve is a real bottleneck in the system (in contrast to compound eyes where it is the optics that limit performance). In humans the optic nerve contains about 1.2 million axons of ganglion cells, compared with about 6.5 million cones and 120 million rods, giving an overall receptor to optic nerve axon ratio of about 100:1. Clearly there is great compression, and it is not hard to guess why. The human optic nerve, 2 mm thick, is flexible enough not to interfere with eye movements; but if each receptor contributed an axon, the nerve would need to be as wide (and so as solid and immobile) as the eye itself.

Amphibious eyes

In all vertebrate groups there are some species that need to see reasonably well in both air and water. Flying fish, mud skippers, most amphibians, turtles, diving birds, seals, and otters all spend part of their lives in each medium. How do they cope with the sudden large changes in optical power when they dive or surface?

We have seen one method already. Seals, many diving birds such as penguins, and some rock-pool fish minimize the problem by having a much flatter cornea than their non-amphibious relatives (Figs 5.6b and 5.8), and thus the cornea has little or no refracting power. The lens has to do nearly all the optical work in both

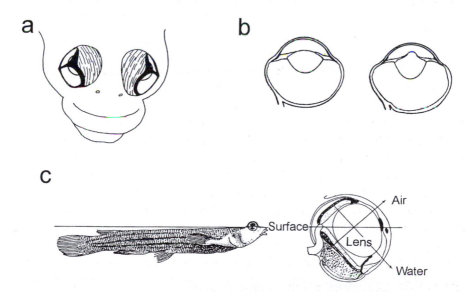

Fig. 5.14 Amphibious eyes. (a) The rock-pool fish *Mnierpes macrocephalus* with flat-faced 'goggles'. (b) Accommodation in the merganser, a diving duck, is achieved by squeezing the lens through the iris to produce a high curvature. (c) The 'four-eyed fish' *Anableps* achieves simultaneous vision in air and water by the use of an ovoid lens with different curvatures on different axes. Various sources.

media, and there is little change in the position of focus on immersion. One problem with a flat cornea is that it tends to restrict the field of view, and also results in serious distortion in the periphery. The rock-pool fishes *Dialommus fuscus* and *Mnierpes macrocephalus* have solved this problem in a particularly interesting way (Fig. 5.14a). They have two flat goggle-like corneas in each eye, making an angle of about 135°. Presumably this both increases the field of view and decreases distortion, although at the price of having a distinct 'join' through the centre of the visual field. A somewhat similar arrangement occurs in the flying fish *Cypselurus heterurus* which has a tent-like cornea consisting of three almost flat triangular facets.

An alternative is to have a focusing mechanism so strong that it can make up the shortfall in optical power. When we dive we lose 40 D (dioptres) and can accommodate by a maximum of 10 D, so we are still left with an unbridgeable 30 D. Certain diving birds, however, have a method of altering the curvature of the front surface of the lens that is much more effective than ours. Birds and reptiles have a muscular iris supported by a ring of bony ossicles around the eye, and they are able to squeeze the lens into the constricted pupil using the powerful ciliary (Brücke's) muscle, creating a very high curvature in the resulting blip (Figs 5.9 and 5.14b). Using this technique mergansers and goldeneyes, both diving ducks, are able to generate 80 and 67 dioptres of extra power respectively, whereas the non-diving wood duck and mallard produce only about 6 and 3 dioptres (Sivak *et al.* 1985). Other diving birds probably use this method of accommodation, as do aquatic turtles and water snakes.

The 'four-eyed fish' *Anableps anableps* from South America has solved the problem of seeing in air and water simultaneously. *Anableps* cruises with half its eye above the surface meniscus, and half below (Fig. 5.14c). It has two pupils, one looking into each medium, and a lens whose shape 'combines an aquatic optical system harmoniously with an aerial one, in a perfectly static situation' (Walls 1942). The compromise is achieved by the ovoid shape of the lens, with its long axis in the direction that looks down into the water. Rays parallel to the axis meet the strongest curvature of the lens, and so are refracted relatively more than rays coming from air, which meet the weaker curvatures of the short axis. The latter rays, however, are also bent by the cornea, so that the total amount of refraction is much the same in the two cases. It seems that this wonderful design is unique.

Invertebrate eyes with corneal optics

It comes as something of a surprise to find that the cornea-lens combination (our kind of eye) is not particularly popular in the animal kingdom. It is necessarily confined to terrestrial animals, and since insects have opted predominantly for compound eyes, that only leaves one major land-living invertebrate group, the spiders, which makes exclusive use of the simple corneal eye system. Some insects also have simple eyes when they are larvae, and adults may also use them as flight-stabilizing devices, in conjunction with the compound eyes.

The eyes of spiders

Spiders and their terrestrial relatives (particularly the phalangids and scorpions) all have eyes of the simple type with the cornea as the main refracting surface. Their distant chelicerate relatives the horseshoe crabs (*Limulus*) have compound eyes, and it may well be that the eyes of modern arachnids are derived from compound eyes by a process of simplification. True spiders usually have eight eyes (sometimes six) and these are of two kinds: the principal eyes which point forwards and the secondary eyes which cover more peripheral fields of view (Fig. 5.15). The two kinds of eye have different embryological origins, and the layout of the receptors is different. In the principal eyes the receptors are similar to those of most other invertebrates. They have a distal segment (nearest to the lens) which bears the photopigment on microvilli, and the cell body and axon are proximal to it (Fig. 5.15). In the secondary eyes, however, it is usually the cell body that is distal, with the microvillous segment forming what is morphologically the first part of the axon (Blest 1985).

Optically the eyes of spiders are very varied. The eyes all quite small, mostly much less than a millimetre across, but this still makes their lenses larger than the facets of compound eyes by an order of magnitude, and so their potential resolution is correspondingly greater. Some have indeed specialized in high resolution, most notably the jumping spiders (Figs 5.16, 5.17, and 5.18). The most impressive

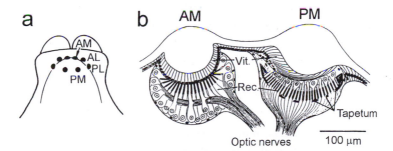

Fig 5.15 Eyes of spiders. (a) Head of the house spider *Tegenaria* showing the four pairs of eyes. The principal eyes (antero-median, AM) have a different structure from the three pairs of secondary eyes (antero-lateral, AL; postero-lateral, PL; postero-median, PM). (b) Details of a principal eye and a secondary eye. In the AM eye the photopigment-containing rhabdoms (shown darker) are distal in the receptor cells (Rec), but in the PM eye the receptor nuclei are distal. Above the retina lie the transparent vitreous cells (Vit). The PM eye has a tapetum, but the AM does not. See also Fig. 6.8. and Plate 3

Fig 5.16 Eyes of *Portia* (*left*), a jumping spider with the highest acuity known in any spider, and *Dinopis*, an 'ogre-faced spider' with the most sensitive eyes. In *Portia* the large eyes (diameter 0.8 mm) are the AM's, in *Dinopis* they are the PM's (diameter 1.3 mm).

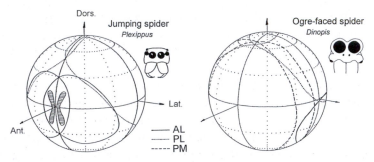

Fig 5.17 Fields of view of jumping spiders and ogre-faced spiders (see Fig. 5.16) showing the way different eyes are used for different purposes. The diagrams show the disposition of eyes on the prosoma (inserts), and the fields of view of the three secondary eyes, which detect movement of the prey. These fields are represented on the surface of a globe with the spider at its centre. In the jumping spiders (left) the AL and PL eyes detect potential prey, which is then identified by the high-resolution principal eyes, whose retinal fields of view are shown here hatched (the retinae of these eyes scan, as indicated on Fig. 5.18a). In the very nocturnal *Dinopis* the PM fields overlap almost completely and presumably pool their signals. *Dinopis* typically hunts from a downward-pointing position above the forest floor, and the AL and PL eyes image the field behind (above) the spider, presumably watching for potential predators. The AM fields in *Dinopis* are similar to the PM's.

of these, *Portia fimbriata*, has an inter-receptor angle of 2.4 arc-minutes; this is only 5 times greater than the human eye, and more than five times smaller than the equivalent angle in the 'best' insect, a dragonfly. Others have modest resolution but enormous light gathering power. Some spiders of the genus *Dinopis*, which catch cockroaches in forests at night, have eyes up to 1.4 mm in diameter, comparable with a small rodent (Figs 5.16 and 5.17). However, the majority of web-building spiders have rather poor eyesight. The principal eyes usually do form

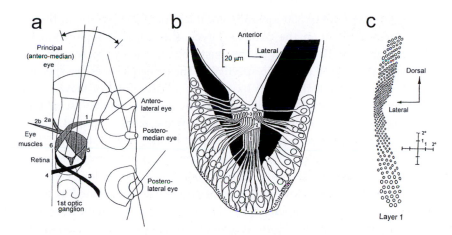

Fig. 5.18 The principal eyes of jumping spiders. (a) Horizontal view of opened prosoma (*right side*) showing the long tubular principal eye and smaller lateral eyes. The principal eye is moved by six muscles in two bands (dorsal muscles in black ventral stippled). These move the eye around three axes, and in the horizontal plane each can move over the arc shown by the thick arrow (see also Chapter 9). (b) Horizontal section of right retina, showing the four tiers of receptors. The tiering is thought to compensate for both focus and chromatic aberration. (c) The distribution of receptors in layer 1. The highest density is in the centre, giving the eye a distinct acute zone. Modified from Land (1985).

low-resolution images, but the secondary eyes have strange unfocused lens–mirror combinations. These are certainly involved in navigation, using the sun and other celestial cues, but exactly how they work is still a mystery (Land 1985).

The largest eyes, and the simplest to understand from an optical point of view, are found in spiders which hunt their prey by sight rather than using webs as traps (Fig. 5.16). These include the families Salticidae (jumping spiders), Lycosidae (wolf spiders), Thomisidae (crab spiders), Sparassidae (huntsmen), and Dinopidae (ogre-faced spiders). Of these the jumping spiders undoubtedly have the most acute vision, and the most sophisticated visual system (Fig. 5.18). They are diurnal hunters that stalk their prey (usually insects) in much the same way that a cat stalks a bird. They turn towards moving objects, directed by the secondary eyes, and then track them using the forward-pointing principal eyes. Oscar Drees studied these spiders in the 1950s, and found that the principal eyes were also responsible for distinguishing between prey and potential mates, and that this judgement was made on the basis of the geometry of the leg pattern of the target animal (Forster 1985). Not surprisingly, in view of the need for fine discrimination, it is the principal eyes that are largest in salticids, with a corneal diameter of 380 μm and a focal length of 767 μm (*F-number* about 2) in a moderately large species, *Phidippus johnsoni* (Land 1985). In *Portia fimbriata*, and probably other species, the focal length is increased by a telephoto arrangement similar to that in hawks (Fig. 5.12; see Williams and McIntyre 1980). By contrast, the postero-lateral (secondary) eyes, which detect movement over a field of 135°, have a corneal diameter of 300 μm and a focal length of 254 μm. In addition, the receptors are narrower in the principal eyes, the smallest separation being 2.0 μm compared with 4.5 μm in the posterolateral eyes, corresponding to angular separations of 9′ and 1°, respectively. The principal eyes are specialized in two other ways. First, they each have a very narrow field of view (about 5° horizontally by 20° vertically), but this is offset by the fact that they 'scan' targets, with a complex pattern of eye movements involving lateral, vertical, and rotational movements of the retinae (see Chapter 9). Unlike vertebrate eyes, the lens itself remains still: it is only the retina that moves. Second, the retina is arranged in four layers, one behind the other. These animals have good colour vision (Plate 4), and this arrangement allows each visual pigment to be situated at the right distance from the lens to compensate for the longitudinal chromatic aberration of the optics. It may also allow objects at different distances to be focused on different layers, and so act in lieu of an active accommodation system.

Of all the eyes considered in this section, the principal eyes of salticids are probably the only ones in which the full optical resolving power of the corneal lens is exploited. Like the human eye, there is a close match between receptor spacing and the diffraction limit, meaning that the eyes are optimized for vision in bright light. Their light-gathering power is correspondingly low, with a calculated sensitivity (Chapter 3) similar to that of diurnal insects. One of the most attractive features of the salticid visual system is its compactness. By confining high resolution to one pair of narrow, long focal length eyes, whilst using much smaller eyes for peripheral vision which requires lower resolution, jumping spiders have saved a great deal of space. If the same eye performed both tasks (as in vertebrates), its volume would be at least ten times greater.

In contrast to the diurnal salticids, wolf spiders are mainly crepuscular or nocturnal hunters. Four of the secondary eyes are much larger than the rest, with corneal diameters of up to 0.4 mm in *Arctosa variana*, and have an *F-number* of 1 or less (Plate 3). The inter-receptor angle is 1–2°, which is similar to the eyes of many insects, and is nowhere near the diffraction limit. Lycosids typically hunt by pouncing on their prey in a single, very rapid combined jump and turn, for which the four posterior eyes are certainly responsible. Thus the main function of these eye is as low-light movement detectors for locating prey, and probably predators as well. Besides having a wide aperture which would help vision at low intensities, the eyes also possess a reflecting tapetum which has the function of doubling the effective length of the receptors (see Chapter 6). It consists of many layers of very thin crystals (probably guanine) which form a long ribbon beneath the receptors (see Chapter 6, Fig. 6.8). Overall, the secondary eyes of lycosids are about 100 times more sensitive (*S*, see Chapter 3) than their salticid equivalents. The principal eyes, however, are relatively small, they lack a tapetum, and are probably not involved in prey capture, although they do seem to be concerned with orientation to the pattern of polarized light in the sky.

The largest eyes of any spider, and probably the largest simple eyes of any land invertebrate, are found in the genus *Dinopis* (Fig. 5.16). As mentioned above, they are nocturnal hunters that ambush insects passing beneath them by pinning them to the substrate with a net of sticky silk – rather like a Roman retarius gladiator. The trigger for this action is visually detected movement. Here the specialized eyes are the postero-medians, with corneal diameters of up to 1.4 mm, a focal length of 0.8 mm, and an extraordinary *F-number* of about 0.6. The severe spherical aberration of an optical system of this size and aperture is counteracted in part by the lenses having a double structure – a low index outer layer surrounding a more dense core – the core itself behaving as a graded-index lens as in fish eyes (Chapter 4). The receptors are also huge, with receptive segments 20 μm wide and 55 μm long during the day, lengthening to twice this in the dark. The other remarkable feature of the receptors is that during the day the microvilli are almost completely resorbed into the proximal part of the cell and reconstituted to fill the rhabdomeres each night. This trick, apparently for protecting the photopigment during the day, is quite common throughout the arthropods. The net effect of these various heroic adaptations is that the sensitivity of these eyes is enormous. Compared with a salticid like *Phidippus*, the sensitivity (measured as the number of photons absorbed per receptor, for a given field luminance; see Chapter 3) is roughly 2000 times greater, although with an inter-receptor angle of about 1.5°, the resolution is about ten times coarser.

Web-spinning spiders have principal eyes that are image-forming, although of relatively low resolution. The secondary eyes, however, are quite different. They typically have weak lenses that form an image well behind the retina, which is itself rather long and thin. However, behind the retina is a tapetum, usually referred to as 'canoe-shaped', which reflects light back and probably focuses it into a very astigmatic line image on the retina. The impression one has is that these eyes are not for 'seeing' in the conventional sense, and indeed there is no indication that movement in their field of view elicits behaviour. They seem to be concerned instead with detecting the direction of the sun and other celestial cues. In some species the principal eyes are responsible for detecting the pattern of polarization in skylight, and from this the sun's direction when it is not visible (Görner and Claas 1985). However, in the wandering spider *Drassodes*, the tapeta of the postero-median eyes have built-in polarizing properties, and these eyes provide the spiders' polarization compass. They use these eyes to find their way back to the nest after foraging trips (Dacke *et al.* 1999).

Corneal eyes in insects

Insect simple eyes, or ocelli, fall into two main groups: the larval eyes of holometabolous insects, and the dorsal ocelli present in most winged adult insects. In both the curved air–tissue cornea interface is the main refracting

a   b   c

Isia   *Euroleon*   *Perga*

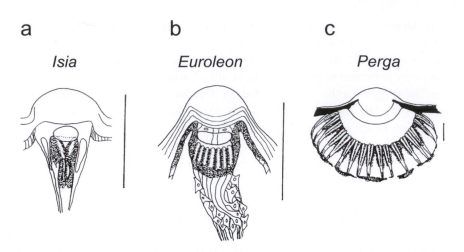

Fig. 5.19 Larval ocelli in insects. (a) The least complex is in the lepidopteran *Isia* where each of the 12 ocelli resembles a single ommatidium from a compound eye. (b) The ant-lion *Euroleon* has six pairs of ocelli which have a more extended retina of 40–50 receptors. (c) The single pair of ocelli of the sawfly *Perga* are more eye-like, with an extended retina and an inter-receptor angle of about 5°. Scale bars are all 0.1 mm. Redrawn from various sources.

surface, although as in vertebrate eyes, a lens of some kind often augments the optical power of the system and aids in the formation of the image.

In insects with a distinct larval stage, the ocelli are the only eyes the larvae possess. They vary greatly in size and complexity. The larvae of flies have no more than a small group of light-sensitive cells on either side of the head. Lepidopteran caterpillars, however, have ocelli with lenses, and a structure resembling that of a single ommatidium from a compound eye. In each ocellus in *Isia*, 7 receptors contribute to a two-tiered rhabdom containing the photopigment (Fig. 5.19a). There seems little possibility of spatial resolution within each ocellus, but as it appears that the fields of view of the 12 ocelli do not overlap, they are capable of providing a 12 'pixel' sampling mosaic of the surroundings. These ocelli do, however, resolve colour; three spectral types of receptor have been found in butterfly larval ocelli.

The ant-lion *Euroleon* (Neuroptera) also has 6 ocelli on each side of the head, borne on a small turret (Fig. 5.19b). Unlike caterpillars, however, each has an extended retina of 40–50 receptors, giving inter-receptor angles ($\Delta\phi$) of 5–10°. Although this resolution is not impressive, it is presumably enough to allow the animals to detect their prey – moving ants at a distance of about 1 cm. Sawflies (Hymenoptera) have larvae with a single pair of ocelli, each with an in-focus retina covering a hemisphere (Fig. 5.19c). The rhabdoms in *Perga* are made up of the contributions from 8 receptors (much as in an ordinary compound eye) and are spaced 20 μm apart, giving an inter-receptor angle of 4–6°. These larvae are vegetarian, and it seems that the main function of the ocelli is to direct the larvae

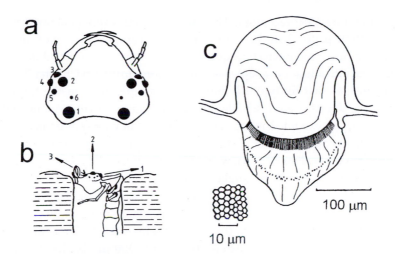

Fig. 5.20 Eyes of tiger beetle larvae (*Cicindela*). These are the largest and best resolving simple eyes in insects, and are used to spot prey (usually ants) which are then caught and pulled down the burrow: (a) head, (b) larva in ambush position, (c) section of largest ocellus showing cornea, lens formed from thickened cuticle, and retina. Inset shows part of retinal mosaic. The inter-receptor angle is about 1.8°. Redrawn from Friederichs (1931).

to their host plants. However, *Perga* larvae will also track moving objects with their head, and defend themselves by spitting regurgitated sap.

The most impressive of all larval ocelli are found in tiger beetles (*Cicindela*). These have a lifestyle similar to ant-lions, ambushing insect prey as they pass their burrows (Fig. 5.20). There are again six ocelli on each side of the head, but two are much larger than the others. The largest has a diameter of 0.2 mm and a

Fig. 5.21 The dorsal ocelli of the locust *Schistocerca gregaria*. (a) Position of the frontal and lateral ocelli on the head. (b) Section of an ocellus, showing the pigmented iris, receptor layer, and layers of neuropil from which a few large axons emerge. The focus positions (light and dark adapted) are very much deeper than the receptors, so this is not an eye that makes use of image detail. (c) The fields of view of the three ocelli, showing how they straddle the horizon. Redrawn from various sources.

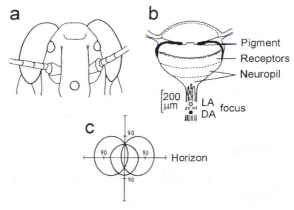

retina containing 6350 receptors. The inter-receptor angle is about 1.8°, comparable with or better than the resolution of the compound eyes of most adult insects. This raises the interesting question as to why the insects did not retain eyes like this into adult life, a topic we will explore further in Chapter 7.

Adult insects that fly typically have three simple eyes on the top of their heads. These dorsal ocelli resemble larval ocelli in possessing a lens and (like sawfly larvae) an extended retina (Fig. 5.21), but they are not embryologically related to the larval eyes. Some dorsal ocelli have tapeta, and some a mobile iris. They each have a wide field of view of 150° or more, and may have as many as 10 000 receptors. So far all this suggests that these are 'good' eyes, like those of hunting spiders. However, there is a problem. Everyone who has tried to get to grips with the optics of these eyes agrees that they are profoundly out of focus, with the retina much too close to the lens (Goodman 1981). For example in the blowfly *Calliphora* the receptors extend from 40 to 100 μm behind the lens, but the focus is at 120 μm. It appears that this is not a mistake; dorsal ocelli are *deliberately* defocused.

What then are they for? A defocused camera is a pretty useless object if detail is to be recorded, so under what circumstances might one *not* want detail? There have been many suggestions over the years, but recent studies mainly support the idea that the ocelli are horizon detectors, involved in enabling an insect to make fast corrections for pitch and roll (Stange 1981). The defocus then makes sense; high spatial frequency clutter such as leaves and branches will be removed, allowing the receptors to respond to changes in the overall distribution of light in the sky. The idea that these ocelli contribute to flight equilibrium is supported by the fact that the receptors converge massively onto a relatively few second-order neurons, and that these project directly into the optomotor system.

Finally, there are a very few examples of simple eyes that seem to be derived from the compound eyes by reduction. The most bizarre are those of male scale insects (*Eriococcus*: Homoptera). A single lens eye occupies the place where each compound eye would have been, and it contains about 500 receptors, giving a quite respectable value for $\Delta\phi$ of 4.7°. Even stranger, the rhabdoms, which in all other insects are composed of microvilli, here contain flattened plates resembling those of vertebrate rods. As pointed out by Paulus (1979): 'The possibility of such modifications demonstrates how easily great changes in organ structure can occur in the evolution of groups'. But it is a good thing that evolution doesn't play tricks like this too often.

Summary

1 Life on land provides animals with a potential new refracting surface – the cornea.

2 For a spherical cornea the nodal point is at the centre of curvature, and with an aqueous fluid behind it the focal length is about four times the radius of curvature.

3 Large eyes are associated either with high resolution or high sensitivity. Nocturnal eyes have larger lenses, relative to the size of the eye, than diurnal eyes.

4 Most land vertebrates have a deformable lens that allows the eye to focus at different distances (accommodation). In humans the optical power (1/focal length) of the cornea is about twice that of the lens.

5 The human cornea has an elliptical profile that corrects for axial spherical aberration.

6 Opening the pupil in man only produces about a tenfold increase in light capture. The pupil's main function is to bring about an optimum balance between sensitivity and resolution. In the gecko, however, the slit pupil can change the brightness of the retinal image by up to 1000 times.

7 Raptorial birds (hawks and eagles) have the highest resolution of any animal, 2–3 times higher than man.

8 The distribution of retinal ganglion cells to some extent reflects an animal's ecology: 'flat-land' animals such as rabbits have a narrow horizontal band of high ganglion cell density (the visual streak).

9 In animals that move between air and water the change in refractive power of the cornea presents a problem. Some have solved it by having a flattened cornea with little power in either medium, others by squeezing the lens into a bony iris to produce a 'blip' of high curvature. The 'four-eyed' fish *Anableps* has a lens with different radii of curvature for looking above and below the meniscus.

10 The spiders are the only other major group whose main organs of sight are single-chambered corneal eyes. The highest resolution is found in jumping spiders (Salticidae) and the highest sensitivity in ogre-faced spiders (Dinopidae). The eight eyes of spiders are of two different structural types; which eyes are used for what purpose varies between different families.

11 Many larval insects have simple corneal eyes, but these are replaced in the adults by compound eyes. Flying insects have unfocussed dorsal ocelli (usually 3) which provide a system of flight stabilization based on horizon detection.

6 | Mirrors in animals

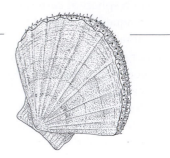

Mirrors seem unlikely things to find in Nature, as living creatures do not produce naturally shiny metals, such as silver or aluminium. Nevertheless, mirrors of various kinds are found performing many functions throughout the animal kingdom. The two most familiar to us are probably the silvery fish that we see in the sea or on the fishmonger's counter, and the eyes of a cat in the headlights of a car. Natural mirrors are not metallic, but are made of multilayers of material with alternating high and low refractive indices (for example air and chitin in insects, water and guanine in fish), and they rely for their effectiveness on interference between the light reflected from the upper and lower surfaces of each layer (Land 1972). We will explore the optical construction of the mirrors later in this chapter, but an interesting and important consequence of the interference is that many natural mirrors are coloured, and this colour can be put to good use in display and camouflage of various kinds. This chapter departs from the general layout of the rest of the book in that it explores the functions of mirrors in structures other than eyes. These all relate to vision, however, and share the same basic mechanism of reflection with mirrors that are found in eyes.

Mirrors in eyes

As with lenses, the value of mirrors as optical components depends on their ability to alter the direction of light rays. The law of reflection states that the angle an incident ray makes with a normal (right angle) to the surface is the same as the angle made by the reflected ray and the normal (Fig. 6.1a). One of the first recorded applications of this capacity to redirect light was Archimedes' scheme to defend Syracuse against the Roman fleet, in which he devised a system of giant mirrors to concentrate the sun's rays on the enemy's sails, to set them alight. The same property also makes it possible to use curved reflecting surfaces in the formation of images. Convex surfaces produce diminished virtual images (i.e. images that cannot be thrown onto a screen, such as the image in a car wing mirror), but concave mirrors can produce real images that can be captured in various ways (Fig. 6.1b and c). Newton was the first to exploit these image-

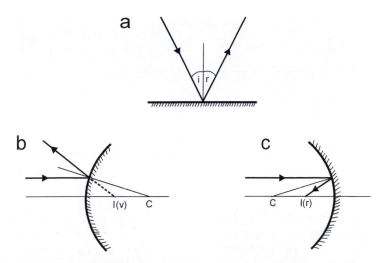

Fig. 6.1 (a) The law of reflection, for a specular (mirror-like) surface. The angle of reflection r equals the angle of incidence i. (b) A convex reflector, such as a car wing mirror, produces an erect virtual image of a distant object at $I(v)$, behind the mirror. $I(v)$ is located half-way between the centre of curvature C, and the mirror surface. (c) A concave mirror, such as a shaving mirror, produces a real inverted image of a distant object at $I(r)$, halfway between the centre of curvature C and the mirror surface.

forming powers in his reflecting telescope of 1671, and since then concave mirrors have been the preferred imaging system for large, high magnification astronomical telescopes, as they are easier to construct than large lenses. Curiously, there is only one good example of an image-forming concave reflector in an eye, and this was not discovered until 1965.

The image-forming reflector in the eye of the scallop

Bivalve molluscs are perhaps not the kind of animal one would look to for optical surprises, or even much in the way of eyesight. However, in this one would be mistaken. A number of genera have evolved optical structures, not to 'see' in any complex sense, but to enable them to detect the approach of predators. Ark shells (*Arca, Pectunculus*) have basic but effective compound eyes in the mantle surrounding the opening of the shells (Nilsson 1994), which evolved quite independently of the more familiar compound eyes of insects and crustaceans (Fig. 7.2c). And scallops of the genus *Pecten* and its close relatives have evolved unique concave reflector eyes for the same purpose (Land 1965).

Scallops have 60–100 small (1 mm) rather beautiful eyes peeping out between the tentacles of the mantle that protects the gape between the two shells (Fig. 6.2a, Plate 1) . Few know of their existence, because this inedible part of the animal is usually thrown away by the fishmonger. A quick look at a section of a

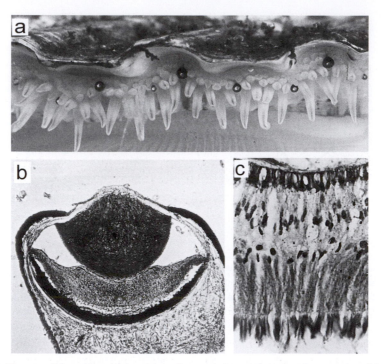

Fig. 6.2 The eye of the scallop. (a) A number of eyes peering out between the tentacles of the mantle which lines each shell. (b) Frozen section of an eye showing the large 'lens', beneath which is a thick retina occupying the whole of the space between the lens and the hemispherical back of the eye, which is lined with a reflecting layer, the argentea. The eye is 1 mm in diameter. (c) Silver-stained section of the retina showing the two photoreceptor layers, distal above and proximal below. The dark structures at the top are the ciliary photoreceptive membranes of the distal cells, and those at the bottom the microvillous membranes of the proximal cells.

scallop's eye (Fig. 6.2b) shows it to be quite like a fish eye. It has a single chamber, so is camera-like rather than compound; and there is a lens of sorts, and behind this a thick two-layered retina filling the space between the lens and the back of the eye (Figs. 6.2b and c). A problem with this fish-eye interpretation of the section is that there is no space between the lens and retina, and had this been a fish eye with a 'Matthiessen' lens there should have been a space of at least 1.5 radii for the converging light rays to focus across (the focal length of a fish lens is ≈ 2.5 lens radii; see Chapter 4). It turns out that the 'lens' is jelly-like, with a low, homogeneous refractive index, and a resulting focal length that would put the focus a long way behind the back of the eye (Fig. 6.4a). There is no way that this could work in the same way as an eye with a fish lens. A more remarkable observation is that when you look into a scallop's eye through a dissecting microscope, you do indeed see an image: an inverted image of yourself looking through a microscope! (Fig. 6.3). It was this observation that finally led to

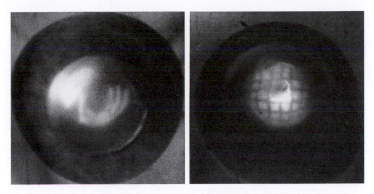

Fig. 6.3 Images in scallops' eyes. *Left*: self-portrait of the author, whose hand is holding the microscope objective used to photograph the eye. *Right*: a grid of 3 mm squares, 15 mm from the eye.

the solution of the optical enigma. The back of the scallop's eye is accurately spherical and lined with a green-reflecting mirror, the 'argentea', so named for its silvery appearance. The image one sees is formed by this reflector, with a small amount of help from the lens (Fig. 6.4a). A calculation of the image position showed that it fell on the part of the retina just below the back surface of the lens; this is the region occupied by the photoreceptive parts of the outer, distal, layer of receptors. Thus this is an eye based on a mirror, not a lens.

Concave mirrors form images on the same side of the reflecting surface as the object (Fig. 6.1c), and if they are spherical they have a focal length (*f*) equal to half the radius of curvature (*r*), i.e.:

$$f = r/2. \tag{6.1}$$

This means that the image of a distant object will be situated half a radius of curvature in front of the mirror. In the scallop it is actually a little nearer to the mirror than this, because the lens has already converged the light slightly. For nearer objects the appropriate equation for working out image position (analogous to eqn 5.5 for refracting surfaces) is:

$$1/v + 1/u = 1/f \tag{6.2}$$

where *u* and *v* are the object and image distances. In this case object, image and focal length are all on the same side of the reflecting surface, so they are all positive in the sign convention. This formula is not of much interest to the scallop, which will be concerned to close its shells when potential predators are a metre or more away, and for an eye this size a metre is practically infinity.

One might ask, why does this eye have a lens at all? It seems that the lens probably does have a function, related to the strange domed shape of its front surface. Just like spherical refracting surfaces, concave mirrors suffer from spher-

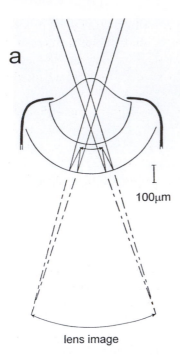
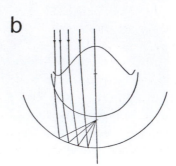

lens image

Fig. 6.4 (a) Image formation in the scallop eye. The lens has very little power, and on its own forms a very deep-lying image. The reflecting argentea forms an image just below the lens, on the region of the cilia of the distal receptor cells (see Fig. 6.2c). b) The probable function of the dome-shaped lens is to correct the spherical aberration of the mirror. The diagram shows a constructed front lens profile that brings all parallel rays to a point focus. Apart from the most peripheral part of the lens, which in life is covered by pigment, the profile is very similar to that of real lenses. The principle is similar to that of a Schmidt corrector plate in a reflecting telescope.

ical aberration (over-focusing of rays at a distance from the axis). This is routinely 'cured' in astromical telescopes with an additional lens called a 'Schmidt corrector plate' whose complex profile manipulates the beam entering the reflector so that there is more focusing power in the centre, near the axis, and less at the periphery. If one works out a profile for a scallop lens that would enable it to make the same kind of correction, it comes to look very much like the profile of the real lenses (Fig. 6.4b), with a high curvature in the centre, flattening towards the periphery (Land 1965).

In 1938, the pioneer neurophysiologist H.K. Hartline had recorded from single fibres in the nerves leaving the two layers of the scallop retina (Fig. 6.2c). He found that the distal layer (where the image is formed) only gave responses to a light going off, and the proximal layer (next to the mirror with no image) to light going on. Much later, in the prime era of electron microscopy in the 1950s and 1960s, it turned out that the photoreceptive regions of cells in the two layers were anatomically very different: the distal cells' photoreceptive structures were made of splayed-out cilia, but the proximal layer had an arrangement of microvilli much like those found in the photoreceptors of other molluscs and in arthropods. Scallops do orient and swim to brighter or darker parts of the environment, presumably using the weakly directional information supplied by the proximal cells, but their more impressive behaviour is to shut when they see a distant object move. This is a behaviour well-known to divers swimming over sandy bottoms. Since distant objects cast no direct shadow, this response must result from changes in the image on the retina itself. What must be happening is that the cells of the distal retina are stimulated to give 'off' responses when either the leading edge of a dark object, or the trailing edge of a light object, crosses the retina in the reflected image.

Other mirror eyes

The mirror design has not been popular. Although it has the advantage of compactness and high light-gathering power, it does have a very serious weakness: it inevitably produces a low-contrast image. The light reaching the image has already passed through the retina unfocused before the mirror returns it as a focused image. This reduces the image contrast to roughly one-half that in an equivalent lens eye, which would be like looking through a fog.

There are many very small eyes that incorporate a mirror, although none of them forms an image of a quality comparable with that in the scallop eye. The cockles (*Cardium*) have eyes that are backed by a mirror, but both their small size and the small number of receptors precludes all but the most rudimentary imaging powers. Some crustaceans, particularly the copepods and ostracods, have small median eyes consisting of three cups, often backed by a reflector, and each containing a handful of receptors. Again, the reflected image may provide some directionality to the fields of view of the receptors, but not very much. The best of these is probably the ostracod *Notodromas*, where ray tracing indicates that a good image is formed on a retina of 18 receptors in each of the lateral cups and 9 in the ventral cup.

Athough most ostracods are tiny bivalved aquatic crustaceans, with an unbroken fossil record going back to the Cambrian, there are some monsters with extraordinarily developed mirror eyes. These are found in the deep-sea genus *Gigantocypris* (1 cm across compared with 1 mm for most of the others). This is Alistair Hardy's description of them:

> The paired eyes have huge metallic-looking reflectors behind them, making them appear like the headlamps of a large car; they look out through glass-like windows in the otherwise orange carapace and no doubt these concave mirrors behind serve instead of a lens in front.
>
> (Hardy 1956)

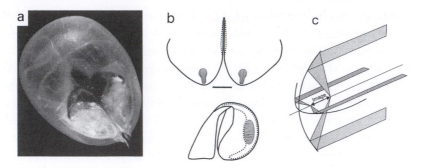

Fig. 6.5 (a) Parabolic reflecting eyes of the deep-sea ostracod *Gigantocypris*. The animal is about 10 mm across (photograph by Dr M.R. Longbottom). (b) Top and side view of the Gigantocypris eye showing the main part of the retina; the hatching shows the approximate orientation of the receptors. The dashed lines enclose much thinner retinal regions. Scale bar 1 mm. (b) Astigmatic line image resulting from the different focal lengths of the parabolic and circular profiles of the reflector, shown in (b). The image lines roughly fit the very long receptors (750 by 25 μm) in the deep region of the main retina.

Hardy made water-colour sketches of *Gigantocypris* and many other deep-sea animals, and he was undoubtedly right about the optical importance of the mirror; but as an imaging system it is certainly very odd. The mirrors are not spherical, but parabolic, and the retinae are not flat sheets as is usual, but condensed into a shape that looks more like a light-bulb than a retina (Fig. 6.5). The curvature of these mirrors in the horizontal and vertical planes is different, which means that the image of a point source will be astigmatic: it will not be a point, but a line at right angles to the mirror (Fig. 6.5c; see Land 1978). The receptors are also elongated in this direction, and so may have some capacity to resolve

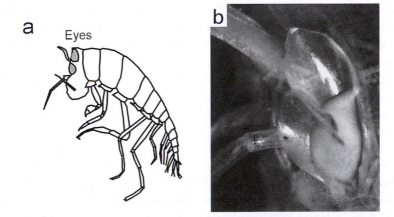

Fig. 6.6 (a) The mid-water hyperiid amphipod *Scypholanceola*. The double eyes are shown stippled. b) Pinna-like reflectors of the double eyes of *Scypholanceola*. The retina is the J-shaped white structure at the base of the reflectors. The height of the whole eye pair is about 2 mm.

these linear images. But everything suggests that the function of these eyes is to concentrate as much light as possible from directions to the left or right of the body axis, rather than producing an image in any conventional sense. At a depth of 1000 m there is no remaining light from the sky (Lythgoe 1979; Denton 1990), so the function of these eyes must be to assist predation by tracking down the luminescent organisms which are common at these depths.

Another enigmatic mirror eye is found in the deep-sea amphipod *Scypholanceola* (Fig. 6.6). This rarely encountered crustacean lives in a similar environment to *Gigantocypris*, and probably uses its eyes for the same purpose. The mirrors are of a very strange shape, looking much more like ears than eyes. There are a pair of these on each side, the upper one is a half-cone rather like a rabbit's pinna, pointing obliquely upwards, and the lower mirror is shorter and more cylindrical, and points forwards. The retinae are open patches of receptors at the base of each reflector. Attempts to model these eyes suggest that the mirrors are efficient light-collectors, capable of indicating at the very least the presence and vague direction of a self-luminous object. As in *Gigantocypris*, however, there is no possibility of an image in the usual sense.

Eyes that use mirrors to produce images are also found in the superposition compound eyes of shrimps, lobsters and their relatives. These will be considered separately in Chapter 8.

Tapeta

A great many eyes have mirrors behind the retina, but unlike the scallop mirror their function is not to form an image. These structures (the correct name is 'tapetum lucidum' meaning silvery carpet) are a common feature of the eyes of vertebrates and arthropods. They are found especially in animals that live in deep water or are active at night. Their function is to reflect the light already focused by the lens, and return it through the retina, giving the retina a second chance of capturing photons missed on the first pass. Because the tapetum is in the focal plane of the lens it has no effect on the optical system of the eye, and the reflected light is returned through the lens as a narrow beam (Fig. 6.7a), visible only from the direction of the original illumination.

Tapeta in vertebrates are made from a wide variety of materials, all having in common a high refractive index. The proportion of the incident light that non-metallic surfaces reflect is closely related to the amount by which their refractive index differs from that of the surroundings; glass, for example, is quite reflective in air, but reflects hardly at all in water. Thus one finds crystals of guanine with a refractive index (n) of 1.83 in the tapeta of the eyes of many fish, riboflavin in the tapeta of bush-babies, and rods containing the zinc salt of cysteine in the tapeta of cats (see below, Fig. 6.10). Many ruminants have a 'tapetum fibrosum' made of collagen, the reflecting properties of which can be appreciated from the white

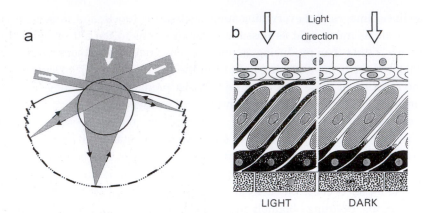

Fig. 6.7 (a) Alignment of the reflecting plates in the tapetum of an elasmobranch fish *Squalus acanthias*. The mirrors are always at right angles to the centre of the image-forming ray bundle, which, for peripheral bundles that only pass through part of the lens, means that they make a steep angle with the retinal surface. Note that ray bundles always leave the eye parallel to the direction that they entered it. (b) The occlusible tapetum in elasmobranchs. In dark adaptation (*right*) the pigment cells are withdrawn but in light adaptation (*left*) they migrate over the surface of the reflecting plates (*hatched*) cutting off the reflection. The retina itself lies above the figure with light coming from the top of the page. Both figures from Nicol (1989).

gleam of muscle tendons. In some teleost fish the tapetum is made up of sub-micrometre spheres of lipid or melanin.

In bright light a tapetum is not needed, and there are a few instances of tapeta that can be occluded as part of the light/dark adaptation process. In many elasmobranch fish, for example, black pigment-containing cells migrate over the surface of the reflecting platelets during the light and retreat during the dark (Fig. 6.7b; see Walls 1941; Nicol 1989). Another feature of elasmobranch tapeta is the way the reflecting platelets are carefully angled, especially at the edge of the retina, to ensure that all untrapped reflected light leaves the eye through the lens (Fig. 6.7a); there is no point in having a tapetum if it scatters light so much that it reduces the contrast of the image.

It might seem an advantage, from a construction point of view, for an eye to have an inverse design. The curious – and seemingly misguided – arrangement in the vertebrate retina, in which the receptors lie at the back of the retina behind the neural layers (Fig. 4.6), makes it possible to 'lay' a single reflecting carpet behind the retina, unencumbered by the need for the axons of the retinal cells to pass through. In support of this, tapeta are certainly uncommon in the eyes of cephalopod molluscs, which have right-way-round (everse) retinae. Spiders, however, have solved this problem. Many crepuscular spiders have tapeta, usually green-reflecting, made in many cases of multilayers of guanine crystals. Some of the most beautiful are in the wolf spiders and their relatives where the tapetum has a 'gridiron' structure, with strips of reflector underlying each row of

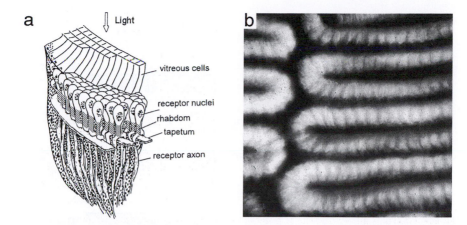

Fig. 6.8 Tapeta in the secondary eyes of lycosid spiders. (a) Diagram of the retina of a lycosid spider, showing how the rhabdomeric (photoreceptive) region of each receptor 'sits' on a strip of reflecting tapetum (modified from Baccetti and Bedini 1964). (b) The tapetum of a lycosid relative (the ctenid *Cupiennius salei*), photographed through the eye's own lens. The tapetal strips are clearly visible, as is the 'join' in the centre of the retina. Each small division of the tapetal strip corresponds to a receptor, and the angle between receptors is about 1°, corresponding to a physical width of about 8 μm. Another lycosid tapetum is shown in Plate 3.

receptors (Fig. 6.8, Plate 3). In other spiders such as the huntsmen (Sparassidae) the receptor axons simply penetrate the continuous tapetum. Tapeta are found in compound eyes as well, especially in the refracting superposition eyes of insects, particularly moths, and the reflecting superposition eyes of decapod crustaceans such as the shrimps, crayfish, and lobsters. In these cases the eyes produce a characteristic eyeglow when viewed from the direction of illumination. The reason is the same as the glow from a cat's or a spider's eye, in spite of the very different basic optical systems of these eyes (see Chapter 8).

Reflecting sunshades

Some animals have mirrors on the outside of the eye, whose function is to keep out direct sunlight that would otherwise scatter within the eye. Some shallow water fishes, such as rays, have a pigmented flap or operculum across the top of the eye, acting as a sun-shade (Plate 1d). Cephalopods such as cuttlefish have an iris with a similar function. However, many other fish have a different solution, which is to use a multilayer mirror instead. These mirrors generally have a green iridescence, and although they appear similar in different species, they are constructed in at least six different ways, implying multiple origins (Lythgoe 1979). The mirrors are organized so that light from above reaches the iridescent layers at a high angle of incidence, which provides a very high reflectance. However, light from objects in the surroundings arrives at close to normal (90°) incidence to the layers, and is only weakly reflected. Thus the fish can see out, but sunlight can't get in.

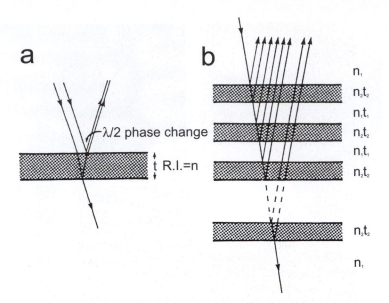

Fig. 6.9 (a) Reflection at a single thin film, for example an oil film or soap bubble. The reflected light from both surfaces will interfere constructively, and the surface appear bright, if the optical thickness of the film (*nt*) is equal to 1/4 of the wavelength of the incident light. (b) In a multilayer structure, maximum constructive interference occurs when all the plates and spaces in the stack have an optical thickness of 1/4 wavelength; i.e. $n_1t_1 = n_2t_2 = \lambda/4$.

The physical optics of animal reflectors

Most of the reflectors found in nature rely on the principle of multilayer interference. Their operation is closely related to the better-known phenomenon of the bright coloured reflections we see in soap bubbles and oil films, and we will deal with these first. In thin films, some light is reflected from the upper surface, and some from the lower surface. If, on re-emerging through the upper surface, the light from the lower surface is in phase with the light reflected directly from the upper surface (that is, the highs and lows of the waves coincide), then the two beams reinforce each other, and the film appears bright; if they are out of phase it appears dark (Fig. 6.9a).

In a soap bubble, the thinnest part of the film is always black. As the thickness increases the film becomes a bright white, then passes through a series of colours (known as Newton's series, but not the same one as the rainbow colours) that start off very vivid and slowly decrease in saturation until the film becomes white again, when it is a few micrometres thick. It is the first white band that is of particular interest from the mirror-making standpoint, as it gives a high reflection over a broad spectral range. It occurs when the *optical* thickness of the film is a quarter wavelength, i.e.

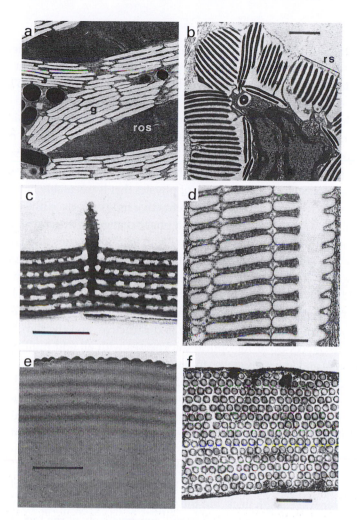

Fig. 6.10 Electron micrographs of natural multilayer mirrors. (a) Tapetum of a bay anchovy *Anchoa mitchilli*, in which a guanine crystal multilayer (g) surronds each rod outer segment (ros). From Nicol *et al*. (1973). (b) Part of a reflecting cell in the skin of *Octopus*, with proteinaceous platelets separated by cytoplasm (rs, reflectosome). From Brocco and Cloney (1980). (c) Section of a wing-scale from the brightly coloured day-flying moth *Urania ripheus* from Madagascar. This is an orange-reflecting scale. The structure is a multilayer of six chitin layers with air spaces between them. See also Plate 2. (d) Chitin–air multilayer in the tapetum behind the receptor cells in the eye of the white peacock butterfly (*Anartia* sp). The tapetum is formed by the extended taenidial ridges of a respiratory trachae-ole, and reflects a bluish colour. From Miller and Bernard (1968). (e) Cornea of a green-reflecting facet from the eye of a horsefly *Hybomitra lasiophthalma* (Diptera: Tabanidae). The distinct layers consist of chitin of different densities, probably indicating different degrees of hydration. From Bernard and Miller (1968). See also Plate 3. (f) Tapetum of a cat, made of a multilayer of rods of a zinc-containing protein. The layers of rods behave in the same way as plates in a conventional multilayer. From Pedler (1963). The scale bar is 1 μm on all the figures; note that it covers 4–5 repeat units of the pattern in all cases.

$$nt = \lambda/4 \tag{6.3}$$

where n is the refractive index, t the actual thickness, and λ the wavelength. The reason for dealing with optical thickness here is that light slows down when the refractive index is high, and the wavelength shortens by a factor of n, so that when distances need to be measured in numbers of wavelengths, this has to be taken into account. Why a quarter wavelength? One might think that it should be half a wavelength, on the grounds that if the light reflected at the lower surface has been twice through the film before emerging, it will have gone through a whole extra wavelength, and so will come out in phase with the light from the top surface. There is, however, a complication. Due to a piece of physics that is resentably hard to understand, light reflected from a low-to-high refractive index interface (the top surface) automatically undergoes a half-wavelength phase change, whereas at a high-to-low interface (the bottom surface) it does not (Fig. 6.9a). This means that constructive interference is achieved if the light from the bottom surface travels a total optical distance of only half a wavelength, meaning that the optical thickness of the film should be $\lambda/4$. When the film becomes vanishingly thin the light from the two surfaces travels the same distance, but the upper surface still imposes its $\lambda/2$ phase change, so the interference is destructive, and the film is black. In fact, high reflectances occur for optical thicknesses with all odd multiples of $\lambda/4$, (i.e. $3\lambda/4, 5\lambda/4$) and low reflectances for even multiples.

Films that reflect in this way are indeed very thin. Blue-green light has a wavelength of 0.5 μm (500 nm), and a quarter of this is 0.125 μm. If the film is mainly water (refractive index 1.33) then the actual thickness is 0.094 μm – several times smaller than the resolution limit of the light microscope.

A single thin film only reflects a few percent of the incident light. However, it is possible to increase this to very close to 100 per cent by adding more films stacked one above the other (Fig. 6.9b). The trick here is to make not only the films themselves a quarter-wavelength thick, but the spaces between them as well, thus ensuring that light from every interface interferes constructively at the top of the stack. Figure 6.10 shows electron micrographs of several different natural multilayers, and the alternating pattern is very clear. In each case the thicknesses of the layers and spaces are all in the region of 0.1 μm, which is what one would expect for quarter-wavelength interference reflectors. A variety of materials is employed: guanine and water in fish (Fig. 6.10a), protein and cytoplasm in *Octopus* (Fig. 6.10b), and chitin and air in many insect structures (Fig. 6.10c and d).

As mentioned earlier, the fraction of light energy (r) reflected at each interface depends on the refractive index difference, according to Fresnel's formula:

$$r = (n_a - n_b)^2/(n_a + n_b)^2 \tag{6.4}$$

where n_a and n_b are the refractive indices of the plates and spaces respectively. If the difference between them is big, then so is the reflectance. As layers are added

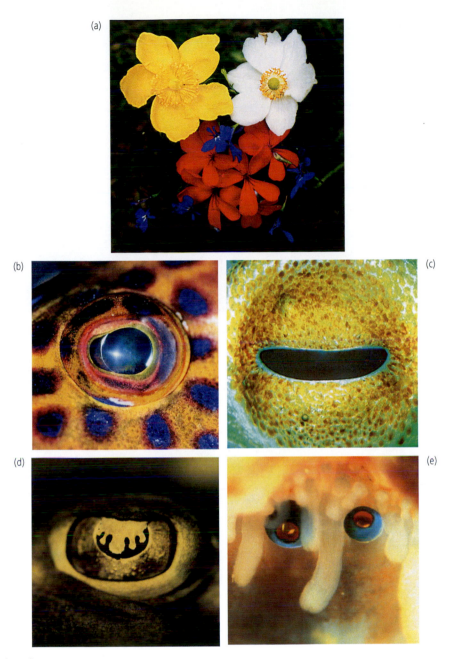

Plate I

a) Four flowers — *Hypericum*, *Anemone*, *Pelargonium* and *Lobelia* — whose spectral reflectance curves are shown in Fig. 2.3b

b) Eye of a coral cod *(Cephalopholis)*, with an aphakic space that permits forward vision. See Fig. 4.7.

c) Eye of an octopus, showing the horizontal slit pupil. See also Fig. 5.11

d) Eye of a shovel-nosed ray *(Aptychotrema rostrata)* with an expanded 'sun-shade' operculum.

e) Two eyes of a scallop *(Pecten)* each about 1mm across. The images of the light sources can be seen in the eye. See Figs. 6.2 to 6.4

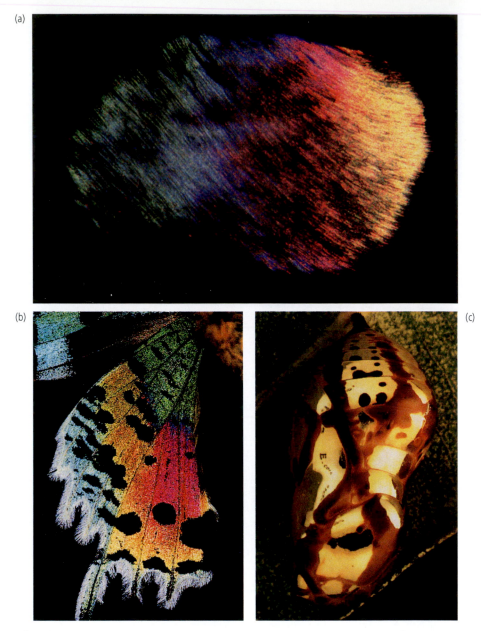

(a)

(b)

(c)

Plate 2

a) Photograph of a single scale from a herring *(Clupea harengus)* showing the different colour zones of the reflecting platelets. Photograph by Eric Denton. See Fig. 6.13d

b) The underside of the hindwing of the Madagascan moth *Urania riphius*. The colours result from constructive interference of light reflected from layers of chitin and air. See Fig. 6.10c

c) The golden pupa of the danaid butterfly *Euploea core*. The quality of the mirror can be judged from the reflection of the animal's name on the left hand side. Photograph by Rudolph Steinbrecht.

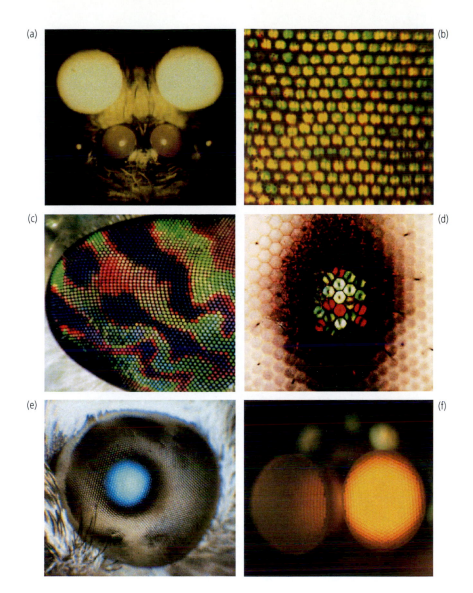

Plate 3

a) Appearance of a lycosid spider when illuminated from the direction of view. The large postero-median eyes glow from light reflected from the tapetum (eye diameter 0.49mm). See Fig. 5.15

b) Retina of the postero-median eye of the lycosid spider *Pardosa prativaga* showing individual receptors on strips of tapetum. Ophthalmoscope photograph by David O'Carroll. See also Fig. 6.8

c) Eye of a horsefly *(Haematopota pluvialis)* with multilayer interference colours. See Fig. 6.10e

d) Red and green reflections from the ommatidia of a butterfly *Heteronympha merope*. The reflections come from multilayer mirrors at the base of each rhabdom See Fig. 6.10d. Waveguide modes originating in the rhabdoms are also visible as lines and dots (Fig 3.7).

e) Blue light reflected from the tapetum of the eye of the hummingbird hawk moth *(Macroglossum)*. The light zone corresponds to the superposition pupil (Fig. 8.6). Photograph by Justin Marshall.

f) Superposition eye of the male mayfly *Centroptilum sp*. The yellow colour is not tapetal, but caused by the scattering of long wavelengths by screening pigment in the retina.

(a)

(b)

Plate 4

a) Male jumping spider *(Habronattus americanus)* showing colourful adornments of the palps and face. See Figs. 5.18 & 9.12. Photograph by Wayne Maddison.

b) Mantis shrimp *(Odontodactylus scyllarus)*, displaying highly coloured appendages. Note the strip through the eye which contains the colour vision system. See Fig. 9.11. Photograph by Justin Marshall.

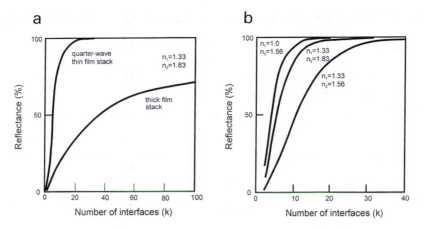

Fig. 6.11 Performance of multilayer mirrors at their wavelength of maximum reflectance. (a) A quarter-wave stack of materials with the refractive indices of guanine and water reaches 90 per cent reflectance after only 10 interfaces, whereas a thick film stack ($nt > 5\lambda$) only reaches 20 per cent. (b) The reflectance for a given number of interfaces in a quarter-wave stack increases with the difference between the refractive indices of the component layers. The three curves correspond to a chitin–air stack (*left*; see Fig. 6.10 c and d), a guanine/water stack (*centre*; see Fig. 6.10a), and a protein or chitin and water stack (*right*; see Fig. 6.10b). Both from Land (1972).

the reflectance of the whole stack (R) increases dramatically. With k interfaces the equivalent formula to 6.4 becomes:

$$R = (n_a{}^k - n_b{}^k)^2 / (n_a{}^k + n_b{}^k)^2. \tag{6.5}$$

The effect, for a guanine–water multilayer with different numbers of interfaces, is shown in Fig. 6.11a. The figure also shows the result for a stack of otherwise similar 'thick films', so much thicker than the wavelength of light that interference no longer occurs (for example rolls of clingfilm, or adhesive tape). Enlisting constructive interference is clearly well worthwhile: the quarter-wavelength stack reaches 99 per cent reflectance after about 20 interfaces (or 10 high index plates), but the 'thick' stack only achieves 30 per cent. The reflectance of quarter-wave stacks of other common combinations of biological common materials is shown in Fig. 6.11b. The number of layers required to achieve a high reflectance depends mainly on the refractive index difference.

Besides a high reflectance, the other important feature of multilayer reflectors is their colour. A multilayer structure is tuned to reflect best at a wavelength four times the optical thickness of the films and spaces. At double this thickness interference is destructive, so no light of twice the preferred wavelength is reflected. Thus these structures are wavelength selective, and so inevitably coloured, which makes them potentially useful in display. It is also possible, by varying the spacing of the plates, to produce white reflectors, which are of more value in various types of camouflage.

Uses of multilayer reflectors in structures other than eyes

Display

Multilayer mirrors are ideal for increasing an animal's conspicuousness in social or sexual contexts. Their silveryness catches the sun, and their colour can be used to specify identity. Birds such as peacocks, birds of paradise, and silver pheasants use coloured multilayer mirrors in display, and more modestly iridescent feathers are part of the plumage of pigeons, starlings and many duck. The structures involved in bird mirrors are often melanin rodlets (which have a refractive index of around 2) embedded in the keratin of the feather. In birds of paradise the rodlets are inflated by nitrogen into flattened plates, with the gas providing a higher refractive index difference.

Multilayer reflectors are also found as display colours in fish. Examples are the coloured adornments of the dragonet (*Callionymus lyra*), and the blue stripe along the body of neon tetra (*Paracheirodon innesi*). The latter has the intriguing property that it can be 'turned off' at night, apparently by a mechanism that decreases the spacing between the guanine platelets. This moves the reflectance peak into the violet-ultraviolet region of the spectrum where it becomes almost invisible (Lythgoe and Shand 1989). There are many other examples of reflectors in fish, and in cephalopods, but the function of most of these is camouflage rather than display (see below).

Amongst insects the iridescence of the wings of some butterflies and moths is particularly striking. The colours of most butterfly wings are pigmentary, but the blues, especially, tend to be structural rather than pigmentary, and depend on constructive interference. Males of the genus *Morpho* have intensely blue wings which were much admired by victorian collectors. These have scales with long ridges, from each side of which protrude plates separated by air spaces providing quarter-wave multilayers. Equally brilliant are the wings of the day-flying Madagascan moth *Urania* (= *Chrysiridia*) *ripheus* (Plate 2). In this species (also much collected) the colours span the range from blue-green, through yellow to red, and then into the next spectrum through purple and blue to green again. Structurally, the chitin layers in the scales increase in optical thickness from 1/4 to 3/4 of a wavelength, whilst the air gaps remain 1/4 wavelength thick (Fig. 6.10c). Other examples of multilayer reflectors in insects include the strikingly coloured corneas of horseflies, deer-flies, and long-legged (Dolichopodid) flies. Here the colours, which disappear after death, are due to alternating chitin layers with different degrees of hydration, and hence different refractive indices (Fig. 6.10e and Plate 3). Their function is unknown. Butterfly eyes also contain reflectors, but here the mirror is a tiny device formed from the chitinous ridges of a tracheole (Fig. 6.10d), and is situated immediately below each rhabdom (photoreceptor structure). The colour, which varies across the eye, can be seen transiently when the eye is illuminated from the direction of viewing. The function of these mirrors, like the tapeta of cats and moths, is to redirect light back through the photoreceptors.

Box 6.1 Spectral reflectance of multilayer mirrors

There is a particularly simple formula for working out the spectral distribution of reflectance of an infinite stack of quarter-wave plates. This will give a useful guide to the way that a stack with a finite number of plates will perform at different wavelengths. The reflectance is given by:

$$R = 1 - \sqrt{(1 - r/\cos^2 \phi)} \qquad (6.6)$$

where r is the reflectance of a single interface, from 6.4, and ϕ is the amount by which the phase of the light is delayed by each plate or space (the 'phase retardation'), given by:

$$\phi = 2nd\ \pi/\lambda \text{ (radians).} \qquad (6.7)$$

The spectral reflectance of infinite multilayers made of different common combinations of materials is shown in Fig. 6.12a. The most obvious difference between the curves is the spectral bandwidth, which depends on r, and hence on the refractive index difference between the high and low index layers. An air–chitin multilayer, of the kind found in iridescent insect wings for example, reflects light over a range of wavelengths nearly three times larger than the water–protein or water–chitin interface found in reflecting surfaces in the skin of cephalopods such as squid. Thus high refractive index differences produce relatively unsaturated colours, which is good for making mirrors.

It is a little more complicated to work out spectral distributions for stacks with a finite number of layers (see Land 1972) but the basic result is that with small numbers of layers the peak reflectance is lower (as in Fig. 6.11b), the bandwidth is somewhat broader, and the central peak has 'sidebands' that get closer together the more layers there are (Fig 6.12b). Two other results are also important. First, the colour of the reflected light changes with the angle of incidence. At normal incidence (at right angles to the surface) the wavelength of maximum reflectance has its highest value, given by eqn. (6.5) for a quarter-wave stack, but as the angle between incident beam and the normal increases this wavelength becomes shorter. Thus iridescent structures with a multilayer construction often change colour with viewing direction. Second, the saturation of the reflected colour varies with the relative optical thicknesses of the high and low refractive index layers. If one is somewhat greater than $\lambda/4$ and the other less than $\lambda/4$, so that the sum comes to $\lambda/2$, the peak reflected wavelength is the same as an all quarter-wave ('ideal') stack, but the bandwidth of the reflected light gets narrower. Because these 'non-ideal' stacks tend to be more highly coloured, they are particularly useful in display.

Box 6.1 Spectral reflectance of multilayer mirrors (*contd.*)

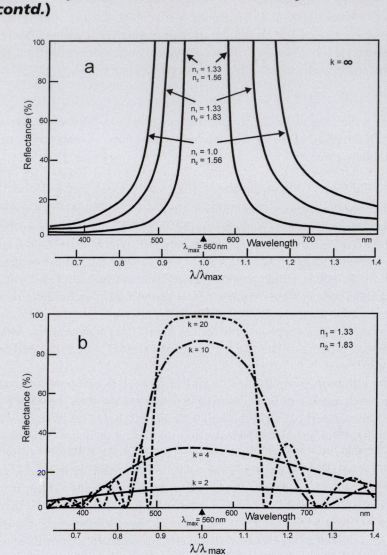

Fig. 6.12 Spectral reflectance distribution of quarter-wave multilayers. (a) Spectral reflectance for multilayers of air–chitin (outer), guanine/water (middle) and protein/water (inner) for stacks with infinite numbers of layers. Note the decrease in bandwidth with decrease in the refractive index difference. The upper ordinate scale assumes that $\lambda_{max} = 4nt = 560$ nm (i.e. yellow-green). The lower ordinate is independent of λ_{max}. (b) Spectral reflectance for guanine–water stacks with small numbers of interfaces. The maximum reflectance rises, the bandwidth decreases, and the number of sidebands increases as the numbers of interfaces (k) increases. Both from Land (1972).

The term 'structural colour' is applied to any system that generates colour using interference of light waves rather than pigment. This includes not only multilayer reflectors but also diffraction gratings and scattering structures, which typically produce rather subdued blue and green colours. Other examples of such structural colours are given in books by D.L. Fox (1953) and H.M. Fox and Vevers (1960), and a review by Parker (2000).

Reflecting camouflage

Mirrors can also have the opposite function to display: that of rendering an animal invisible. In a beautiful series of papers in the 1960s, Denton and Nicol showed how the silvery sides of fish provide a form of camouflage which, in the context of the open ocean, makes their bodies very difficult to see (Denton and Nicol 1965; Denton 1970). The principle is simple, and relies on the fact that as sunlight penetrates below the sea surface, wave refraction and scattering diffuse the light so that it becomes nearly symmetrical around a vertical axis (Fig. 6.13a). At any particular angle to the vertical light coming from, say, the north is similar to that from the south, and this is more or less independent of the sun's angle relative to the surface. In this situation, a vertical plane mirror becomes invisible, because the light reflected from it is identical in intensity to the light that would have passed through it (Fig. 6.13b). The efficacy of this can be judged from the photograph of a silvery fish (Fig. 6.14), which is almost invisible until it tilts out of the vertical and reflects light from above. Lythgoe (1979, Fig. 7.1) shows a similar photograph in which the most visible feature of the fish is the black pupil of the eye. This cannot be disguised because it must absorb light if vision is to work. Divers occasionally report being passed by shoals of black dots, and regret the excesses of the previous evening.

To make this reflecting strategy work, fish have had to solve two problems. First, fish are not flat-sided – they bulge – and the camouflage trick will not work unless the mirror is fairly accurately flat. Second, the mirrors must be white, not coloured, as a simple quarter-wave stack made of guanine and water would be. Denton and Nicol showed that the bulge problem was dealt with in a most ingenious way. Although the sides of most fish are convex, the reflecting platelets in the scales themselves are not parallel to the body surface, but are tilted so that they are aligned much more closely with the vertical (Fig. 6.13c). Thus the side of the fish behaves optically as a plane vertical mirror, independent of its real shape. As mentioned earlier, a white reflector can be made by varying the thickness of the layers within a mulitilayer stack. In fish like the herring and sprat this is done in a very neat way. Within each scale the colour reflected by the multilayer varies, so that approximately a third of the scale is blue-green, a third red-purple, and a third orange-yellow (Plate 2). The scales overlie each other rather like roof tiles, so that they are three deep at any one point, with the differently

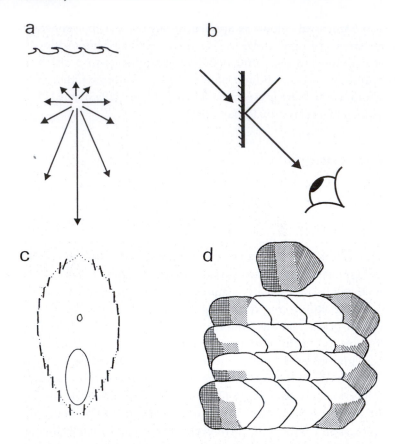

Fig. 6.13 Reflecting camouflage in the sea. (a) At a depth greater than a few tens of metres the distribution of light around the vertical becomes symmetrical. (b) A vertical plane mirror becomes invisible in the sea because the light reflected at any angle has the same intensity as the light that would have passed through. (c) The orientation in the vertical plane of reflecting platelets around the body of a herring. Note that they conform much more closely to the vertical than to the body surface. (d) The overlap of reflecting scales in the herring. Each scale has regions each reflecting a different 1/3 of the spectrum, and when three overlap each other the reflection is white. All based on Denton (1970).

coloured multilayers one on top of the other (Fig. 6.13d; see Denton and Nicol 1965). Since multilayer reflectors transmit what they do not reflect, each scale is able to reflect its own one-third of the spectrum unhindered in transmission by the other two scales, and the net result is an impressively white reflection, as the display on the fish counter attests. It is only when silvery fish become damaged, or are simply not as fresh as they might be, that they lose scales and start to become colourful.

An additional component of the camouflage strategy of many mid-water fish, cephalopods and crustaceans is 'counter-illumination' in which rows of down-

Fig. 6.14 A silvery fish (a permit, *Trachynotus falcatus*) oriented vertically in the sea (*top*) and tilted (*bottom*) so that it reflects light from above. (Photographs by Justin Marshall).

ward-pointing luminescent photophores (usually involving multilayer reflectors) are used to disguise the silhouette of the animal when viewed from directly below (Herring 1994).

On land the mirror strategy usually won't work because light is much more directional. There is one setting, however, that has a light environment a little like the ocean. This is the deep forest. Here light is diffuse, and the background in one direction looks much like that in any other. Pupae of certain danaine butterflies, for example *Euploea core* from Sri Lanka, have evolved brilliant gold-reflecting multilayer cuticles (Plate 2), whose surfaces reflect the details of the surrounding forest undergrowth (Steinbrecht *et al.* 1985). This is perfect camouflage; the intensity and texture match the surroundings, and invisibility is assured.

Summary

1 A small number of eyes employ concave mirrors, rather than lenses, as image-forming structures. The most impressive of these is in the scallop Pecten, where the image in each of the 60–100 eyes provides a means of detecting movement.

2 The deep-sea ostracod *Gigantocypris* has a pair of eyes with parabolic reflectors. These provide high sensitivity but poor resolution.

3 Reflecting tapeta that do not form images, but which double the effective light path through the retina, are common in vertebrate eyes (e.g. cat) and also in compound eyes of some insects and crustaceans.

4 All animal mirrors employ the principle of multilayer interference. Light is reflected from a number of layers in a stack, and if the interfaces are separated by a quarter-wavelength, or an odd multiple of this, constructive interference occurs and a high reflectance is produced.

5 Materials involved in biological multilayers include guanine and cytoplasm (fish scales) and chitin and air (insect wings). The highest reflection is produced when the refractive index difference is high.

6 Because the reflectance of a multilayer is a function of wavelength, most biological reflectors are coloured. This makes them useful in display, for example in the iridescent feathers of some birds, and the wings of some butterflies and moths.

7 The special light conditions in the ocean make it possible to use mirrors as an effective form of camouflage. The silvery scales disguise the sides of the fish, by reflecting light with the same brightness as the background.

7 | Apposition compound eyes

Origins

Judging from the numbers of individuals that possess them, compound eyes are by far the most popular devices for imaging an animal's surroundings. Built as convex structures around the outside of an animal's head, they are fundamentally different from the concave structure of single chamber eyes. In spite of this major topological difference, however, the jobs of the two kinds of structure are the same – to break up the incoming light according to its direction of origin (Fig. 7.1). The other great difference between the two kinds of eye is, of course, that compound eyes employ multiple optical systems compared with the single optical system of so-called 'simple' eyes. This does not necessarily mean that compound eyes form multiple images, however. In *apposition* eyes, such as those of most diurnal insects, each of the lenses does form a tiny image (although this is not what the animal actually sees). But in *superposition* eyes, more commonly found in nocturnal insects and deep-water crustaceans, the lenses (or sometimes mirrors) operate in concert to form a single deep-lying image. Because the optical mechanisms involved are very different from each other we have split our discussion of compound eyes into two: this chapter deals with apposition eyes and their variants, and the next with superposition eyes.

Compound eyes first appear at the time of the Cambrian radiation event (see Chapter 1). Several of the more peculiar animals of the Burgess Shale, such as *Anomalocaris*, had large convex eyes, and from their shape these must have been compound eyes, even though the facet structure has not been preserved (Conway-Morris 1998). The arthropod sub-phyla Crustacea and Chelicerata, which go back to the Cambrian, were equipped mainly with compound eyes, as were the first insects which appeared later, in the Devonian. An animal almost unchanged from that early period is the horseshoe crab, *Limulus* (Chelicerata), whose famous compound eyes provided visual physiologists with one of their best preparations from the 1930s onwards. More recent chelicerate groups, such as the scorpions and spiders are thought to have converted the ancestral compound eyes to simple eyes by some process of coalescence. Some of the best pre-

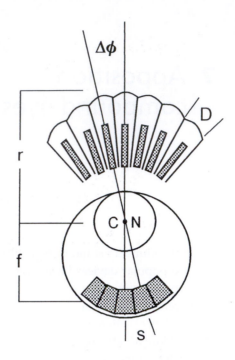

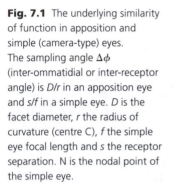

Fig. 7.1 The underlying similarity of function in apposition and simple (camera-type) eyes. The sampling angle $\Delta\phi$ (inter-ommatidial or inter-receptor angle) is D/r in an apposition eye and s/f in a simple eye. D is the facet diameter, r the radius of curvature (centre C), f the simple eye focal length and s the receptor separation. N is the nodal point of the simple eye.

served fossil eyes of any animal group are those of the Trilobites, whose history begins in the Cambrian and ends in the Permian, 300 million years later. The calcite in the exoskeletons of these arthropods has preserved not just the external structure of these eyes, but to some extent the optics too, allowing us a tantalizing glimpse into visual systems half a billion years old (Levi-Setti 1993). Trilobite eyes are discussed later in the chapter (see Fig. 7.21).

All the animals mentioned so far belong to the Arthropoda, and they probably originated from a worm-like ancestor that already possessed a rudimentary compound eye – possibly a loose collection of eyespots like those found in some modern annelids and myriapods. There are, however, two small unrelated groups that have evolved compound eyes independently of the arthropod lineage (Fig. 7.2). The ark shells (*Arca* and *Pectunculus*) are bivalve molluscs with an array of compound and simple eye structures around the edge of the mantle. They fulfil much the same function as the mirror eyes of scallops (Chapter 6), namely as 'burglar alarms' for the detection of moving predators (Nilsson 1994). Unlike arthropod compound eyes these are lensless, with the acceptance angles of the receptors constrained simply by the shadowing effect of the pigmented tubes around them. Sabellid tubeworms are annelids that filter-feed with tentacles which project from a tube half buried in mud, and like the ark shells they need early-warning of approaching predators. In *Branchiomma* the two compound eyes are borne at the tips of specially modified tentacles, and although they do have lenses they are more like the eyes of Arca than those of arthropods

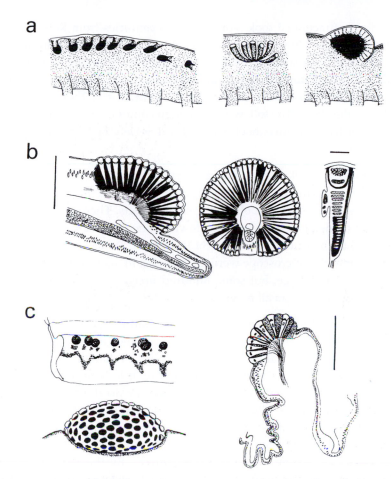

Fig. 7.2 (a) Primitive compound eyes in sabellid worms. From left: *Hypsicomus, Protula,* and *Sabella*. (b) Well-developed compound eye in the sabellid *Branchiomma*. Longitudinal and transverse sections (scale 100 μm) and a single element, in which the receptive part is a stack of ciliary discs (scale 10 μm). (c) Mantle eyes of the bivalve mollusc *Arca*. Section through a single eye on the right (scale 100 μm). Land (1981) from various sources.

(Nilsson 1994). Some starfish, too, have rather loosely organized compound eye-like structures at the ends of the arms, but they seem to be more a collection of small eye-cups of a rather basic kind, rather than a single eye. Their ability to resolve an image is uncertain.

A little history: apposition and neural superposition

The facets of compound eyes of insects are just too small to be resolved with the naked eye, and it required the invention of the microscope in the seventeenth century before they could be properly depicted. The process of working out how compound eyes functioned took more than two centuries from Robert Hooke's first drawing of 'The Grey Drone Fly' (probably a male horse-fly) in his *Micrographia* of 1665, to the essentially modern account by Sigmund Exner in 1891. The first person to look through the optical array of an insect eye was Antoni van Leeuwenhoek, and his observations caused a controversy that was not fully resolved until the 1960s. The following quotation comes from Wehner (1981) and is from a letter from Leeuwenhoek to the Royal Society of London, which was published in 1695.

> Last summer I looked at an insect's cornea through my microscope. The cornea was mounted at some larger distance from the objective as it was usually done when observing small objects. Then I moved the burning flame of a candle up and down at such a distance from the cornea that the candle shed its light through it. What I observed by looking into the microscope were the inverted images of the burning flame: not one image, but some hundred images. As small as they were, I could see them all moving.

Evidently, each facet of the eye (at least in apposition eyes) does produce an inverted image (see Chapter 8, Fig. 8.2), even though the geometry of the eye as a whole dictates that the overall image is erect (Fig. 7.1). What, then does the insect see? Do the receptors (typically eight) beneath each lens resolve the inverted images (as Hollywood would like us to believe), or do they just indicate the average intensity across the field of view of the ommatidium? (An ommatidium is the 'unit' of a compound eye, consisting of the lens, receptors, and associated structures. See Fig. 7.3.)

Remarkably, the answer depends on the animal. By the 1870s histological studies had shown that in most apposition eyes the eight receptor cells in each

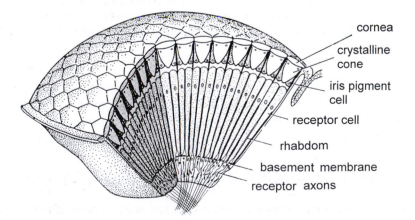

Fig. 7.3 Basic structure of an apposition eye, showing its construction from ommatidial elements. Modified from Duke-Elder (1958; Fig. 134)

apposition

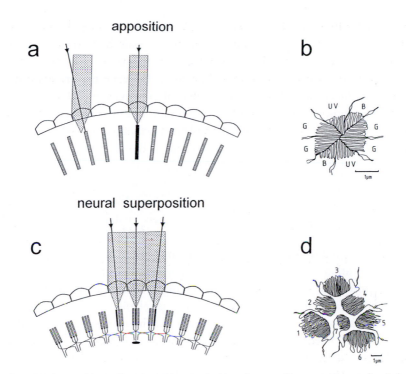

Fig. 7.4 Optical comparison of an apposition eye (a, b) and a neural superposition eye (c, d). In an apposition eye each rhabdom (hatched) views light from a slightly different direction (*arrows*), and the rhabdoms (b), although made up from eight receptors, have a fused structure that acts as a single light-guide. UV, B, and G indicate the receptor elements that respond to ultraviolet, blue and green in an ommatidium from the eye of a worker bee. In neural superposition eyes, light from a single direction is imaged onto different rhabdomeres in adjacent ommatidia (c). The axons from all receptors imaging the same point collect together in the first synaptic layer (the lamina) so that here the image has the same structure as in an ordinary apposition eye. The section (d) shows the arrangement of the separated rhabdomeres in an ommatidium from a fly (see also Fig 7.10c). The six outer rhabdomeres (1–6) all send axons to different adjacent laminar 'cartridges', as in (c). The central pair (7 overlying 8) bypass the lamina and go straight to the next ganglion, the medulla.

ommatidium contribute to a single radial structure, known as a rhabdom (Greek for rod; Figs 7.3 and 7.4). Much later, in the 1950's, this material was found to be made up of photoreceptive membrane covering large numbers of long narrow microvilli, but even by the time that Exner wrote his monograph in 1891 it was clear that the rhabdom was the structure sensitive to light. Optically, each ommatidium works as follows. The inverted image that Leeuwenhoek saw is focused onto the distal tip of the rhabdom. Having a slightly higher refractive index than its surroundings, the rhabdom behaves as a light guide, so that the light that enters its distal tip travels down the structure, trapped by total internal reflection. Any spatial information in the image that enters the rhabdom tip is lost, scrambled by the multiple reflections within the light guide, so that the rhabdom itself acts as a photocell that averages all the light that enters it. Its field of view is defined, in geometric terms, by the angle that the tip subtends at the nodal point of the corneal lens (see Fig. 7.6), and in a typical apposition eye this *acceptance angle* ($\Delta\rho$) is approximately the same as the angle between the ommatidial axes (the *inter-ommatidial angle*, $\Delta\phi$). Thus the field of view of one rhabdom abuts (or 'apposes', hence the name) the field of its neighbour, thus producing an overall erect image made up of a mosaic of adjacent fields of view.

Although the eight receptors that contribute to the rhabdom share the same visual field, that does not mean that they supply the same information. The labels UV, B, and G on the cross-section of a bee rhabdom in Fig. 7.4b indicate the regions of the spectrum that the cells respond to best. Most insects have trichromatic colour vision, just as we do, although their visible spectrum is shifted towards shorter wavelengths compared with ours (Menzel 1979; Chittka 1996). Some butterflies and dragonflies have four-colour vision, and so does the water-flea *Daphnia* – rather implausibly given that it only has 22 ommatidia. Most other crustaceans are di- or tri-chromatic. An amazing exception is the mantis shrimp *Odontodactylus* (Stomatopoda) which has 12 visual pigments in a specialized band across the eye (see Chapter 9 and Plate 4). The second feature of the bee rhabdom (Fig. 7.4b) is that the microvilli making up the structure are arranged in orthogonal sets. It has been known since the work of Karl von Frisch in the 1940s that bees can navigate using the pattern of polarized light in the sky. This capacity arises from the way the photoreceptor molecules are arranged on the microvilli (see Chapter 2). A geometric consequence of the cylindrical shape of the microvilli is that there will be twice as many light-sensitive chromophore groups of the rhodopsin molecules aligned parallel to the long axis of each microvillus than at right angles to it. This in turn means that the receptors respond best to light polarized parallel to this axis. In fact bees use a special dorsal region of the eye (the POL area) to analyse sky polarization; in the rest of the eye the receptors are twisted to abolish polarization sensitivity, so that it does not interfere with colour vision (Rossel 1989; Wehner 1987). Polarization vision is also used by some insects, such as the water bug *Notonecta*, to detect water surfaces, which polarize light strongly.

The description of apposition optics given above holds for most diurnal insects and crustaceans (bees, grasshoppers, water fleas, crabs, etc.) but it does not apply to the true (two-winged) flies. Ever since 1879, when Grenacher observed that the receptors in fly ommatidia have separate photoreceptive structures (rhabdomeres) that do not contribute to a common rhabdom, there had been suspicions that flies might actually be resolving the Leeuwenhoek images. In the focal plane of the lens of a fly ommatidium the distal tips of the rhabdomeres are separated from each other and form a characteristic pattern (Fig. 7.4d, see also Fig. 7.10c) which resolves the image into 7 parts (there are 8 receptors, but the central pair lie one above the other). This raises the obvious question: how are these 7-pixel inverted images welded together to form the overall erect image, if indeed that is what occurs? Kuno Kirschfeld finally solved this conundrum in 1967. It turns out that the angle between the fields of view of adjacent rhabdomeres *within* an ommatidium (about 1.5° in a blowfly) is identical to the angle *between* neighbouring ommatidial axes. Furthermore, the fields of each of the 6 peripheral rhabdomeres in one fly ommatidium are aligned, in the space around the fly, with the field of the central rhabdomere of one of the neighbouring ommatidia (Fig. 7.4c). Thus each point in space is viewed by 7 rhabdomeres in 7 adjacent ommatidia. What does this complicated and seemingly redundant arrangement achieve? To answer this it is necessary to know what happens to the signals from the 7 receptors that view the same point, and that turns out to be the most astonishing part of the story. Beneath each ommatidium the emerging receptor axon bundle undergoes a 180° twist before the individual neurons disperse to nearby regions of the first optic ganglion (the lamina) that correspond to the adjacent ommatidia. The net result of this impressive feat of neural knitting (indicated in Fig. 7.4c) is that all the axons that 'look at' the same point in space finish up making connections with the same cells in the lamina. Thus, as far as the lamina is concerned, the image is exactly the same as it would be in a conventional apposition eye, except that the signal, in terms of photon captures, is seven times stronger. One advantage of the extra signal is that it provides flies with a short period at dawn and dusk when they can see well, but the eyesight of their predators and competitors is less sensitive and so less effective at detecting small objects.

Kirschfeld called this arrangement 'neural superposition', because, as in optical superposition (Chapter 8), the contributions of a number of ommatidia are superimposed in the final image. One might ask: could the signal not have been made stronger simply by increasing the diameter of the rhabdom in a conventional apposition eye? Indeed it could, but that would mean increasing the rhabdom acceptance angle ($\Delta\rho$) at the same time, which in turn would mean a loss of resolution for the eye as a whole. The beauty of the fly solution, and undoubtedly the reason why it evolved, is that it involves no increase in acceptance angle, provided the rhabdomeres are properly aligned. There are strong hints that something like neural superposition occurs in other insect groups (some beetles, earwigs, water bugs and craneflies) but it is only in the advanced flies that the perfect nearest-neighbours arrangement is known to be achieved (Nilsson and Ro 1994).

Basic optics

Most of the optical theory given in Chapters 3 and 5 applies to apposition compound eyes, but there are some differences from camera-type eyes. In this section we outline the major points again, and make comparisons with other types of eye.

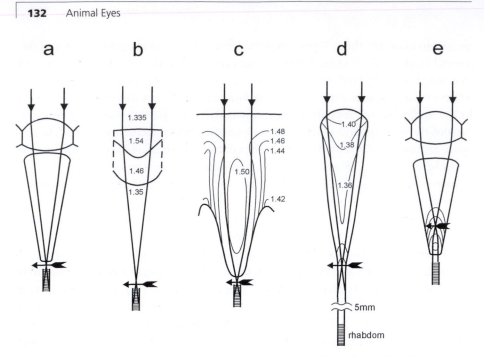

Fig. 7.5 Five mechanisms of image formation in apposition eyes. (a) Corneal lens (bee, fly). (b) Multi-surface lens (water-bugs). (c) Graded-index lens-cylinder (*Limulus*). (d) Lens-cylinder with light-guide (*Phronima*, Amphipoda). (e) Lens/lens-cylinder afocal combination (butterflies). Details in text.

Imaging mechanisms

The structures that form the images in the ommatidia of apposition eyes are quite varied (Fig. 7.5). In terrestrial insects, as in terrestrial vertebrates (Chapter 5), the simplest way to produce an image is to make the cornea curved (Fig. 7.5a). Ordinary spherical-surface optics then apply (see Chapter 5), and an image is formed about 4 radii of curvature behind the front face. In aquatic insects such as the water bug *Notonecta* the external surface of the cornea has little power, because of the reduction in refractive index difference (Fig. 7.5b). It is augmented by two other surfaces, the rear of the lens, and an unusually curved interface in the centre of the lens whose function may be to correct spherical aberration, as has been proposed for some trilobite eyes (Levi-Setti 1993).

The horseshoe crab *Limulus* lives mainly in the sea, but comes ashore to lay eggs. It has a flat cornea, but behind this lie a series of inward-pointing conical projections which form images at their proximal tips (Fig. 7.5c). Exner (1891) worked out that, in the absence of any optically useful interfaces, these structures must operate as graded-index devices, forming images by continuous ray-bending much as occurs in the spherical Matthiessen lenses of fish (Chapter 4). He called these structures 'lens cylinders', and his assumption that they must have internal refractive index gradients has been repeatedly confirmed in recent years. Particularly interesting lens cylinders are found in hyperiid amphipods (deep-sea cousins of the more familiar sand-hoppers). The most impressive of these, *Phronima*, has a double eye in which the upper part covers the dorsal surface of the head, and the lower part is a small ventrally situated tear-drop-like structure (Fig. 7.19). The upper eye has graded index lenses not unlike those of *Limulus*, but instead of imaging directly onto the rhabdom tip they focus into the mouth of a long light guide, 15μm wide and with a refractive index of 1.39, which conveys the light from the image 5 mm to the retina, situated ventrally next to the retina of the lower eye (Fig. 7.5d). The function of this peculiar arrangement seems to be camouflage, to keep the eye as transparent as possible (Nilsson 1989), and other mid-water hyperiids have similar arrangements.

The eyes of butterflies, which resemble ordinary apposition eyes in nearly all respects, have a optical system that is subtly different from the arrangement in Fig. 7.5a. Instead of forming an image at the rhabdom tip, as in the eye of a bee or locust, the image lies within the crystalline cone. The proximal part of the cone contains a very powerful lens cylinder which makes the focused light parallel again, so that it reaches the rhabdom as a beam that just fits the rhabdom (Fig. 7.5e). This arrangement, known as afocal apposition because there is no external focus, has much in common with the superposition optical system of moths (Chapter 8), to which butterflies are closely related.

Resolution

As discussed in Chapter 3, for any eye the resolution of the image seen by the brain is determined by the sampling frequency of the eye (v_s) and by the optical quality, represented by the spatial cut-off frequency (v_{co}). In an apposition eye the sampling unit is the rhabdom in a single ommatidium. Although the eight receptors that contribute to each rhabdom usually have different spectral and polarization responses, they all share a common field of view. Thus it is the angle between ommatidia ($\Delta\phi$) that determines how the overall image is sampled (Fig. 7.1), where $v_s = 1/(2\Delta\phi)$. [In a hexagonal array the exact definition of $\Delta\phi$ can become quite complicated (see Fig. 7.15), but for now we take it to mean the average of the angle measured along each of the three axes of the array.] In the central region of a bee eye, $\Delta\phi$ is about 1.7°.

The neural superposition eyes of dipterans have an additional constraint, namely that the separation of the tips of the rhabdomeres must match the inter-ommatidial angle, i.e. $\Delta\phi = s/f$, where s is the separation and f the focal length of the facet lens. If $\Delta\phi$ is 2° (0.035 radians) and f is 70 μm, then the tip separation must be 2.4 μm. This doesn't leave a great deal of room. Because narrow light-guides, such as rhabdomeres, tend to be 'leaky', with a substantial fraction of the light energy outside the guide itself (Fig. 3.7), there needs to be an adequate gap between one rhab-

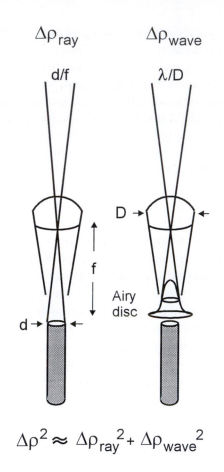

Fig. 7.6 The acceptance angle ($\Delta\rho$) of an ommatidium results from a combination of the Airy diffraction pattern (point-spread function) given by λ/D (*right*), and the geometrical angular width of the rhabdom (d/f) at the nodal point of the lens (*left*). An approximation to $\Delta\rho$ is given by $\sqrt{[(\lambda/D)^2 + (d/f)^2]}$.

$$\Delta\rho^2 \approx \Delta\rho_{ray}^2 + \Delta\rho_{wave}^2$$

domere and the next to prevent 'cross-talk'. In flies there is a 1 μm gap between adjacent rhabdomeres, which means that the rhabdomeres themselves must be very narrow. They have a distal tip diameter which is also about 1 μm, making them amongst the narrowest photoreceptors in any animal. In most other respects, however, neural superposition eyes are optically similar to other apposition eyes.

As in the human eye (Chapter 5) one would expect that apposition eyes would show a rough match between the inter-ommatidial angle and the acceptance angle ($\Delta\rho$) of a single rhabdom, the argument being that no individual rhabdom can resolve detail finer than $\Delta\rho$, so there is no point spacing the directions of view of ommatidia closer than this angle. Just as in other eyes, geometrical (ray) optics and physical (wave) optics both contribute to $\Delta\rho$ (Fig. 7.6). Geometrically $\Delta\rho_{ray}$ is the angle subtended by the rhabdom tip at the nodal point of the facet lens, i.e. the rhabdom diameter divided by the focal length (d/f radians). Typical values (for a bee) are 2 μm for d and 60 μm for f, which makes $\Delta\rho_{ray}$ 0.033 radians, or 1.9°. In wave optics the limit to image quality is set by diffraction, specifically by the angle subtended by the Airy disc, and this (see Chapter 3) is given by λ/D radians. If the wavelength (λ) is 0.5 μm and the facet diameter (D) is 25 μm, then $\Delta\rho_{wave}$ is 0.02 radians, or 1.1°. To obtain the final value for $\Delta\rho$, $\Delta\rho_{ray}$ and $\Delta\rho_{wave}$ have to be combined, and unfortunately the proper way of doing this (convolution, taking the waveguide properties of the rhabdom into account, see van Hateren 1989) is very complicated. Snyder (1979) provides a simple approximation:

$$\Delta\rho^2 = \Delta\rho_{ray}^2 + \Delta\rho_{wave}^2. \tag{7.1}$$

This is adequate for most purposes but tends to overestimate $\Delta\rho$. Using this approximation, $\Delta\rho$ for the bee data is 2.2°, somewhat larger than $\Delta\phi$.

Using the argument that works for humans, namely that the optical cut-off frequency ($1/\Delta\rho$) should match the sampling frequency ($1/2\Delta\phi$), we would expect the ratio of $\Delta\rho$ to $\Delta\phi$ to be 1:2. In fact, it is only 1:1.3 in the bee, and this is fairly typical of diurnal insects (Land 1997). It implies that apposition eyes tend to under-sample the image slightly, or put another way, they operate at levels of contrast in the image considerably higher than those experienced by the human eye at its resolution limit.

Although diffraction imposes severe limitations on the performance of compound eyes (see next section), in many other respects they are excellent instruments. Optical defects other than diffraction tend to have a greater impact on resolution as eyes get bigger (Land 1981). The very short focal length of the facet lenses of compound eyes, 100 μm or less, ensures that such defects as spherical and chromatic aberration, which are troublesome in camera-type eyes, are negligible in compound eyes. Similarly the depth of field is enormous, extending to infinity from as close as an insect ever needs to see.

Diffraction and eye size

In a short and remarkable paper on 'Insect sight and the defining power of compound eyes', published over a century ago, Henry Mallock, an optical instrument maker, described insect vision in these terms:

> The best of the eyes … would give a picture about as good as if executed in rather coarse wool-work and viewed at a distance of a foot.
>
> (Mallock, 1894)

Why is insect vision so poor? The problem, as Mallock recognized for the first time, is diffraction. Compound eyes have very small lenses compared with the lenses of camera-type eyes. As we have seen, a 25 μm diameter facet produces a diffraction blur circle (Airy disc) that is just over 1° wide in angular terms, and cannot resolve spatial frequences higher than 1 cycle per degree. 1° is about the size of a finger-nail at arm's length, so one can imagine a bee's world made up of

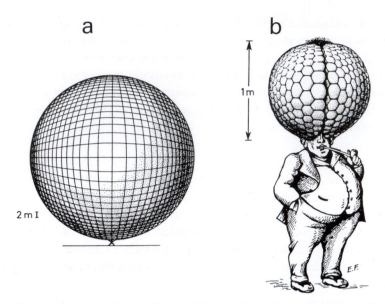

Fig. 7.7 The sizes of compound eyes with human-like resolution. *Left*: a compound eye with 1 minute resolution everywhere. *Right*: compound eye with 1 minute resolution in the fovea, but falling off with eccentricity as in the human eye. Both figures from Kirschfeld (1976).

pixels of about that size. In terms of the acuity of our own eyes (about 60 c/deg), this is not very good at all.

Mallock's paper goes on to discuss what a compound eye with human resolution would look like, and he came to the astonishing conclusion that it would need to be more than 20 metres in diameter – bigger than a house (Fig. 7.7a). The reason for this is clear. The human eye achieves 60 c/deg resolution by having a daylight pupil diameter of 2 mm, 80 times the diameter of a bee lens. For a bee to have the same resolution, diffraction requires that all its lenses would need to have this diameter, and to exploit all the detail in the scene they would need to be spaced at 0.5 arc-min angular intervals, the same as the receptors in our fovea. In a spherical eye, the inter-ommatidial angle ($\Delta\phi$) is the angle subtended by one lens diameter at the centre of the eye (D/r radians, where r is the eye radius; Fig. 7.1), which gives $r = D/\Delta\phi$. With $\Delta\phi = 0.5$ minutes of arc, (0.000145 radians), and $D = 2$ mm, the radius of curvature will be 13.8 m, and the diameter twice this.

Kirschfeld (1976) has pointed out that this calculation is a little unfair. Resolution in the human eye falls off dramatically away from the fovea, to a tenth of its maximum value at 20° from the fovea, and even less further out. Taking this into account the 'human' compound eye can be shrunk in size considerably, to an irreducable 1 metre diameter (Fig. 7.7b). This still looks silly, however, and would certainly be hard to fly with. The serious point is that because of diffraction compound eyes are stuck up an evolutionary blind alley. For a single-lens camera-type eye only one lens needs to be made larger to improve resolution, but for a compound eye all have to be enlarged and the numbers have to increase correspondingly. The net result is that the size of camera eyes increases linearly with resolution, but compound eye size increases as the square of resolution. Dragonflies seem to approach the limit of what it is possible. Their eyes are 8 mm or more in diameter, have up to 30 000 facets each, and resolve about 0.25° in their most acute region. This is still poor compared with what is achievable by any camera-type eye of the same diameter.

The outcome of this discussion is that it is very hard for an apposition eye to improve its resolution – it simply gets too big. Space is thus at a premium; a little extra resolution here must be bought by a bit less there, and for this reason the different visual priorities of arthropods with different life styles show up in the distribution of inter-ommatidial angles, and often facet sizes, across the eye. We will return to this point later when discussing the various ecological adaptations of compound eyes.

Sensitivity

The sensitivity of an apposition eye is calculated in the same way as for a camera-type eye. The formula for sensitivity was derived in Chapter 3:

$$S = 0.62\, D^2 \Delta\rho^2 \tag{7.2}$$

where D is the lens diameter, and $\Delta\rho$ the rhabdom acceptance angle (Fig. 7.6) (we ignore the effect of receptor length here). Although D is roughly 100 times greater in a human eye than in a bee ommatidium, $\Delta\rho$ is about 100 times smaller (approximately 0.015° compared with 1.5°), so that the value of S is very similar in the two species. Thus the range of illumination conditions over which an insect with an apposition eye can operate is similar to that of a mammal using its cone system. Mammals can also see at much lower intensities, by pooling the responses of rods over quite large retinal areas. This process involves a serious loss of resolution, and although pooling may occur in some arthropods (Warrant *et al.* 1996) the numbers of ommatidia involved are relatively small.

When discussing sensitivity, 'adaptation' can have two meanings. Different eyes may be adapted in the evolutionary sense to work permanently in conditions of high or low illumination: night or day, deep-sea or surface. Alternatively, the same eye can be said to be light- or dark-adapted via reversible and temporary changes in its optical anatomy. In both cases eqn 7.2 is the key to interpreting changes and differences. Figure 7.8 shows ommatidia from eyes of two crustaceans, a shallow-water blue crab *Callinectes* and a deep sea isopod *Cirolana*, to illustrate the extent of permanent adaptation to two extremes of lighting conditions. Values for D and $\Delta\rho$ derived from the figure indicate that *Cirolana* is

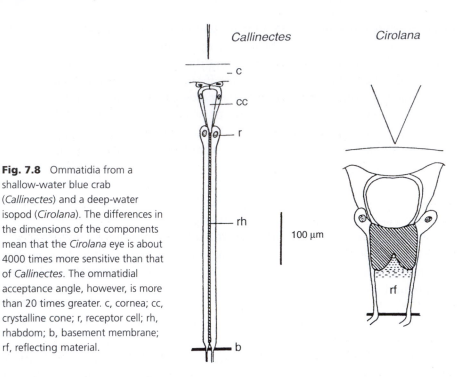

Fig. 7.8 Ommatidia from a shallow-water blue crab (*Callinectes*) and a deep-water isopod (*Cirolana*). The differences in the dimensions of the components mean that the *Cirolana* eye is about 4000 times more sensitive than that of *Callinectes*. The ommatidial acceptance angle, however, is more than 20 times greater. c, cornea; cc, crystalline cone; r, receptor cell; rh, rhabdom; b, basement membrane; rf, reflecting material.

about 4000 times more sensitive than *Callinectes*, the main effect coming from the wide rhabdom acceptance angle in *Cirolana* which results from the massive rhabdom diameter. The cost of high sensitivity, in terms of decreased resolution is very great: $\Delta\rho$ is about 47° in *Cirolana* compared with 2° in Callinectes. As with other types of eye, sensitivity and resolution are in conflict, and to excel in both requires an eye of prohibitive size.

Light and dark adaptation

Temporary light and dark adaptation mechanisms take a number of forms in apposition eyes (see Autrum 1981). Some are illustrated in Fig. 7.9, and include the following. (a) An iris mechanism just above the distal tip of the rhabdom which restricts the effective value of $\Delta\rho$ in eqn (7.2.) In the case of crane-flies (Tipulidae), which have an arrangement of 6 outer and 2 central rhabdomeres, the iris cuts off the outer six in the light leaving only the central pair. (b) A 'longitudinal pupil' consisting of large numbers of very small pigment granules which move into the region immediately around the rhabdom in the light and withdraw in the dark. The main effect of this is to absorb the wave guided light that travels just outside the rhabdom (Fig. 3.7). This is replaced from light within the rhabdom, and this is absorbed in turn, so that light is progressively 'bled' out

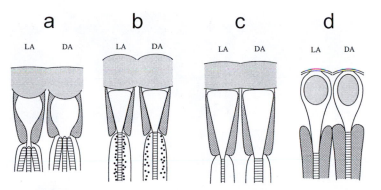

Fig. 7.9 Dark and light adaptation in apposition ommatidia. (a) Variable pupil in front of a rhabdom (tipulid flies, water bugs). (b) Radial migration of pigment granules in the retinula cells (flies, butterflies). (c) Changes in rhabdom size and shape (crabs, orthopteran insects). (d) Changes in lens focal length (*Artemia*). Nilsson (1989).

of the rhabdom. This mechanism is particularly important in higher Diptera (house-flies, etc.) and in butterflies, and it can work in a matter of seconds. (c) The rhabdom dimensions may themselves change, usually over a period of hours. This mechanism may involve the resynthesis of photoreceptive membrane in the dark, and its sequestration in the light. (d) Other photomechanical changes include movements of the rhabdom towards or away from the lens, and in the case of the small crustacean *Artemia* there is a change in the focal length of the lens itself, which shortens in the dark and so increases $\Delta\rho$ in eqn (7.2). In addition to these changes there are electrical and enzymatic changes in the receptors themselves, that alter the gain of transduction and increase response time in the dark.

The pseudopupil

Before describing the different ways that the resolution of apposition eyes is matched to behaviour and ecology, it will be helpful to discuss an optical curiosity known as the pseudopupil. This optical phenomenon provides us with a powerful and non-invasive technique for studying the way resolution varies across a compound eye. Insects and crustaceans with light-coloured apposition eyes have an easily visible dark spot which has the alarming property that it moves across the eye as the observer rotates around the animal (Fig. 7.10). It seems as though one is being watched. In fact this is a passive optical phenomenon that has nothing to do with the visual process itself. Whatever the background colour of the eye, the region that images the observer must look dark because it absorbs photons from the observer's direction. The dark spot (the pseudopupil) moves with the viewer because different parts of the eye image

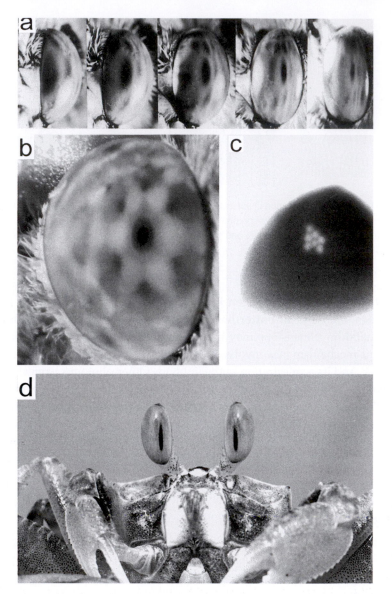

Fig. 7.10 Appearance of pseudopupils in eyes of insects and a crustacean. (a) The Australian bee *Amegilla* photographed at 20° intervals around the eye (front at left). The pseudopupil appears to move round the eye as the head is rotated. It is elongated dorso-ventrally, implying greater vertical resolution than horizontal, and it becomes narrower with increasing angle from the front meaning that the horizontal resolution decreases from front to side. (b) Appearance of a butterfly eye (*Junonia villida*) with a complex pseudopupil that shows the arrangement of the different types of pigment cells in the plane of focus of the ommatidial lens system (see Figs 7.5e and 7.11). The darkest dot in the centre is the image of the rhabdom itself. (c) Antidromic deep pseudopupil of the fly *Drosophila melanogaster*. The pseudopupil has the same geometry as the 7-rhabdomere structure in Fig. 7.4d, with the same characteristic asymmetry. (d) Extreme vertical elongation of the pseudopupil in the ghost crab *Ocypode*, related to increased vertical resolution around the horizon.

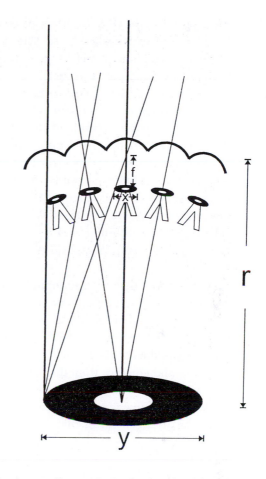

Fig. 7.11 Explanation of the pseudopupil. Seen from outside, rays emerging from the centre of each of several ommatidia appear to come from a single enlarged ommatidium situated at the centre of curvature of the eye (local radius r). Other regions of the ommatidial focal plane superimpose in the same deep region to give a pattern of width y that resembles that in each ommatidium (width x). A good example of such a pattern is seen in Fig. 7.10b.

different directions in space; it is almost disappointingly simple. Often, however, the pseudopupil is more than a dark spot, and has a pattern to it which on careful inspection turns out to be an enlarged and slightly fuzzy image of the various structures around the rhabdom tip, in the focal plane of each facet lens.

Figure 7.11 is an attempt to explain this. When one views the eye from a close distance, rays (dark ones!) joining the tips of several rhabdoms to the eye or microscope form a cone that appears to originate in the centre of the eye. Similarly, rays leaving points just to the left of each rhabdom seem to come from a point just to the left of the centre of curvature. Thus the 'deep pseudopupil' has the same geometry as the structures in each focal plane, but is composed of the superimposed contributions from many ommatidia. Using the principle of similar triangles, it can be seen that the deep pseudopupil is enlarged relative to the original structures in the ommatidium by a factor of r/f (eye radius divided by facet-lens focal length).

We can learn a great deal from the pseudopupil (Stavenga 1979). Its form reveals structures in the ommatidium, without recourse to histology. For example in the butterfly pseudopupil (Fig. 7.10b) the pattern of pigment cells surrounding the rhabdom tip is impressively displayed. The overall shape of the pseudopupil (whether it is elongated in one direction or another) indicates asymmetries in the eye's resolution. For example, a vertically elongated pseudopupil generally implies a larger radius of curvature for the vertical than the horizontal plane, which in turn means that more ommatidia sample a given vertical angle than a horizontal one. An extreme example of this is seen in ocypodid crabs (ghost crabs, fiddler crabs, Fig. 7.10d), but a similar pattern is found in many insects (Fig. 7.10a). However, perhaps the most useful feature of the pseudopupil is that one can use it to measure inter-ommatidial angles. If one rotates an insect's head through a degrees, and the pseudopupil appears to move across b facets, then the inter-ommatidial angle is a/b degrees. Variations in inter-ommatidial angle in different planes, and in different regions of the eye, can be mapped in this way, revealing how the eye is organized to make the most of its limited acuity (Horridge 1978).

Often eyes are so dark that no pseudopupil is visible. This is the case in many dipteran and hymenopteran insects. A useful technique is then to try to obtain an *antidromic* pseudopupil. This method, pioneered by Nicholas Franceschini, involves shining a light up through the base of the head, illuminating the proximal ends of the rhabdoms or rhabdomeres. If it works, light passes up the light-guiding structures and emerges from their distal tips to give a luminous pattern that has the same geometry as a conventional (or *orthodromic*) pseudopupil, and can be exploited in the same way. In flies this works particularly well (Fig. 7.10c), and shows up the arrangement of rhabdomeres in the focal plane very clearly (Franceschini, 1975).

Ecological variations in apposition design

As we have seen, the optical design of apposition eyes means that there is no spare room on the head surface, and what there is needs to be used as efficiently as possible. This in turn means that the disposition of the optical axes of the ommatidia in space will be matched to the visual needs of the animal – to its ecology (Land 1989, 1999). A survey of the apposition eyes of insects and crustaceans leads to the conclusion that there are three main patterns of acuity distribution that one can identify fairly easily. These are: (a) patterns related to the velocity flow-field encountered in forward locomotion, especially flight; (b) 'acute zones' associated with predation or sex, these zones sometimes developing into separate components of a double eye; and (c) horizontal strips of high resolution in animals living in environments such as water surfaces and sand-flats where almost all important activity takes place around the horizon. Figure 7.12 illustrates these situations.

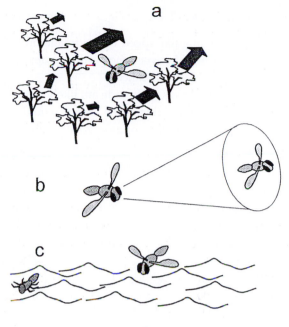

Fig. 7.12 Three reasons why there should be differences in resolution across compound eyes. (a) During forward flight close to vegetation the relative angular velocities of objects are greatest at the side (*heavy arrows*). The resulting blur is matched by lowered horizontal resolution. (b) Pursuit behaviour requires increased resolution, usually in the dorso-frontal quadrant. (c) Close to flat surfaces most objects of interest are near the equator of the eye, where there is often a strip of high vertical resolution.

The forward flight pattern

When an animal is moving through the world, the objects in it appear to move backwards across the eye, in a pattern that has become known as a velocity flow-field (discussed in more detail in Chapter 9). Objects to the sides move faster than those in front, and there is a point in the direction of the animal's travel (the 'focus of expansion') where there is no image motion (see Fig. 9.7). Objects further away move more slowly than near objects. The geometry of image motion is shown in Fig. 7.13, and the relation between motion, position, and distance is summed up in the following expression:

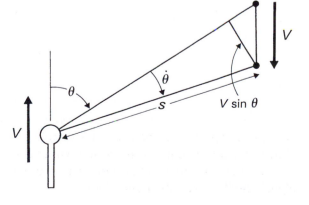

Fig. 7.13 The relation of retinal angular velocity (flow) to distance and angle.
An animal moving with velocity V will see an object at distance s moving across its retina at an angular velocity of $\dot{\theta}$ when the angle from the front is θ. From the geometry of the figure, $\dot{\theta}$ is given by $(V \sin \theta)/s$.

$$\dot{\theta} = V \sin\theta / s \qquad\qquad (7.3)$$

where $\dot{\theta}$ is the angular speed of the image on the retina, V the animal's actual velocity, θ the angle between the particular object and the animal's heading direction, and s the object's distance.

Clearly, near objects to the side are likely to move so fast across the retina as to cause blurring, and if this is the case it would be economical to employ fewer receptors there, as high resolution is not usable. A butterfly or bee spends much time flying past foliage, and reasonable values for s might be 0.5 m, with V about 2 m.s⁻¹. If θ is 90°, then eqn (7.3) gives the speed across the retina as 4 radians or

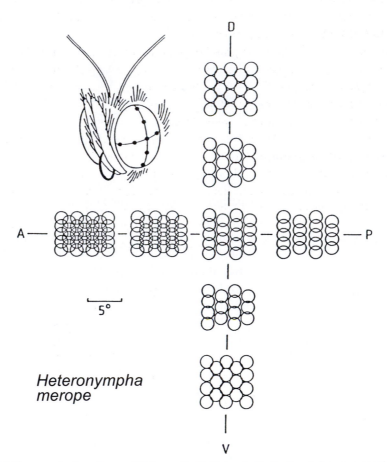

Fig. 7.14 The organization of ommatidial receptive fields in a butterfly. The circles represent the acceptance angles ($\Delta\rho$) of a light-adapted butterfly. In *Heteronympha* this has an almost constant value of 1.9° across the eye. Two trends are clear. Going from anterior (A) to posterior (P) the axes of the ommatidia separate, so that $\Delta\phi_h$ (see Fig. 7.15) has roughly doubled by 120° from the front. From dorsal (D) to ventral (V) the ommatidial axes come together in the region of the eye's equator, and then separate again. These two trends are seen in most flying insects.

229° per second. A typical response time of a light-adapted insect photoreceptor is 10 ms, which means that in one response time the image will have moved $229 \times 0.01°$, giving a blur streak about 2.3° long. It follows that there is little point in having lateral-pointing receptors closer together than 2–3°, however good the resolution at the front of the eye may be. This seems to be borne out in practice. In the butterfly *Heteronympha merope*, for example the horizontal inter-ommatidial angle decreases from 1.4° in front to 2.6° at the side (Fig. 7.14).

In describing acute zones it is helpful to indicate how densely the ommatidial array samples different regions of the surroundings (Fig. 7.15a). The measure adopted here is the number of ommatidial axes per square degree. This is easily calculated from the partial inter-ommatidial angles $\Delta\phi_h$ and $\Delta\phi_v$ as defined in Fig. 7.15b (see Stavenga 1979). The axis density is then $1/(2\Delta\phi_h\Delta\phi_v)$, or $1/(\sqrt{3}\Delta\phi^2/2)$ if the array is symmetrical.

Bees, butterflies, and acridid grasshoppers are flying insects, and their eyes all show decreasing horizontal inter-ommatidial angles from front to rear, consistent with these ideas (Fig. 7.10a). Non-flying insects, for example many tettigonid grasshoppers, have more or less spherical eyes, without this gradient. In all the flying groups there is another, separate gradient of vertical inter-ommatidial angles (Figs 7.14 and 7.16a); they are smallest around the eye's equator, and increase towards both dorsal and ventral poles. This results in a band around the equator with enhanced vertical acuity (Horridge 1978; Land 1999). The most likely reason for this vertical gradient is that the region around the eye's equator contains the highest density of information important to the animal, especially if it is an insect that feeds on flowers. The need for higher acuity is obviously greatest in that part of the field. Vertebrates that live in open landscapes (rabbits, chee-

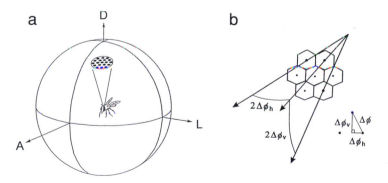

Fig. 7.15 (a) Representation of resolution as the number of ommatidial axes (black dots) in a given (conical) solid angle in the space around an insect. In Figs 7.16 and 7.17 the contours represent equal numbers of ommatidial axes per square degree. (b) Convention adopted by Stavenga (1979) for describing inter-ommatidial angles in an array where vertical and horizontal angles differ. This array is of the bee/grasshopper type (hexagons on their points) but the system applies just as well to dipteran fly/butterfly arrays (hexagons on their sides).

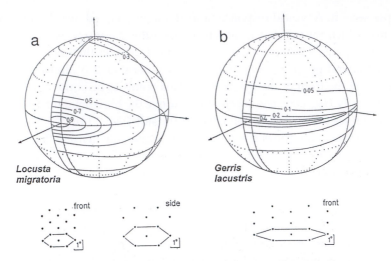

Fig. 7.16 Distribution of resolution, expressed as ommatidial axes per square degree (see Fig. 7.15), for a typical flying insect (locust) and a 'flatland' insect that lives on the surface film (water strider). In *Locusta* the resolution decreases from front to back, and from the equator to the dorsal and ventral poles, as in butterflies (Fig. 7.14). In *Gerris* the equatorial streak is very pronounced, and as the insert shows it is due to an extreme distortion of the lattice of axes, giving much greater vertical resolution than horizontal. The pattern of facets on the eye itself does not show this distortion.

tahs) show a related pattern of increased acuity, but here it takes the form of elongated regions of high ganglion cell density corresponding to the horizon, and known as 'visual streaks' (Fig. 5.13, see also Hughes 1977). More extreme versions of the vertical gradient are found in animals from really flat environments such as beaches and water surfaces (Fig. 7.16b). These are discussed in more detail later.

The combined effects of these two gradients on the overall density of ommatidial axes is shown for a locust in Fig. 7.16a. Worker honey bees, butterflies (Fig. 7.14) and female blowflies (*Calliphora*) show a similar pattern, although in male flies and drone honey bees, this pattern is distorted to give a more pronounced acute zone concerned with mate capture (see Fig. 7.17 a and b).

Acute zones concerned with prey capture and mating

Many insects and crustaceans have a forward or upward-pointing region of high acuity, related either to the capture of other insect prey, or to the pursuit in flight of females by males. Where both sexes have the specialization (mantids, dragonflies, robber-flies, hyperiid amphipods) predation is the reason, but more commonly it is only the male that has the acute zone (simuliid midges, hoverflies, mayflies, drone bees) indicating a role in sexual pursuit. The acute zones vary considerably. In male houseflies and blowflies (Fig. 7.17a) they may

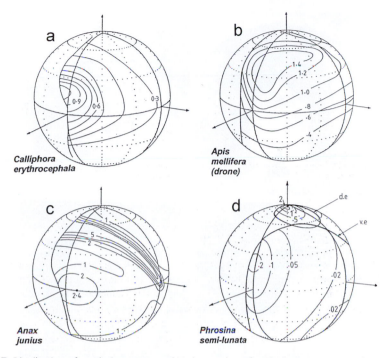

Fig. 7.17 Distribution of resolution, expressed as the number of optical axes per square degree (see Fig. 7.15), in the eyes of four arthropods with acute zones concerned with capture (Fig. 7.12b). (a) Male blowfly. Here the frontal acute zone appears to be an enhancement of the 'forward flight' pattern (cf. Fig. 7.16a) which is present in both sexes. (b) The pattern in drone bees is quite different from workers, with a dorso-frontal acute zone in which the axis density is more than three times that in the worker eye. (c) Some dragonflies have a weak acute zone in the direction of flight, and another band across the fronto-dorsal region, in which the axis density (5 deg^{-2}) is higher than in any other insect (see Fig. 7.18c). (d) Like a dragonfly, the mid-water amphipod *Phrosina* has high upward resolution. This is in the dorsal part of a divided eye (Fig. 7.19c), which has a very small field compared with the smaller-faceted ventral eye.

involve little more than a local increase in the acuity of the 'forward flight' acute zone common to both sexes (see above). However, in other insects and many crustaceans the acute zone may be in a separate eye, as is the case with the dorsal eyes of male bibionid flies (Fig. 7.18b), or the upper eyes of hyperiid amphipods (Fig. 7.19). In these more extreme double eyes, the upward-pointing part is often specialized for detecting other small animals against the sky, or – in the sea – against the residual downwelling daylight.

Good examples of forward-directed acute zones are found in the praying mantids, predators in which both sexes ambush prey. The eyes have large, binocularly overlapping acute zones which are used to centre potential prey before it is struck with the spiked forelegs. Mantids provide the only known example in insects where prey distance is determined by binocular triangulation.

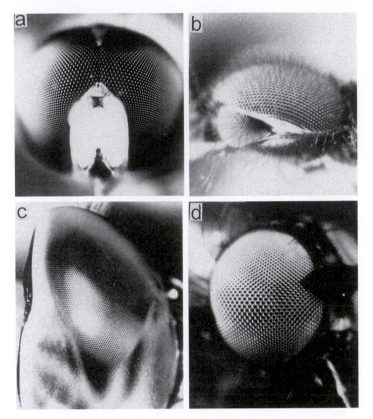

Fig. 7.18 Eyes in which variations in resolution are reflected in the sizes of the facets. (a) Male *Syritta pipiens*, a small hoverfly in which the male has a region of enlarged facets in the dorso-frontal region. It uses this to 'shadow' females which have no such acute zone (see Fig. 9.5). (b) A male bibionid fly (*Dilophus* sp) with a divided eye in which the upper part provides higher resolution for sighting and tracking females against the sky. The females lack the upper eye altogether. (c) Upper part of the eye of the dragonfly *Aeschna multicolor*, showing the wedge of enlarged facets corresponding to the line of enhanced resolution seen in Fig. 7.17c. This is present in both sexes. (Photograph by Truman Sherk). (d) Anempidid fly (*Hilara* sp). Like *Gerris* (Fig. 7.16b) this is a water surface insect and has a region of large facets and high resolution around the equator. (Photograph by Jochen Zeil.)

The inter-ommatidial angle ($\Delta\phi$) in *Tenodera australasiae* varies from 0.6° in the acute zone centre, to 2.5° laterally. Facet diameters decrease from 50 μm in the acute zone to 35 μm peripherally, but this is less of a decrease than would be expected from diffraction considerations alone (Rossel 1979). Amongst crustaceans there are few known examples of frontal acute zones concerned with predation, but perhaps the best documented is in the carnivorous water flea *Polyphemus*, which uses its single fused compound eye to locate and track swimming prey. The *Polyphemus* eye has 130 ommatidia, and includes a distinct acute zone of 22 ommatidia, where inter-ommatidial angles are as small as 2°, which is remarkable in a 0.2 mm eye. The structure of the eye also indicates the use of polarized light in prey capture.

In many male dipteran flies an acute zone is associated with sexual pursuit, and is typically situated 20 to 30° above the flight direction (Figs. 7.17a). In *Calliphora* it is characterized by a low value for $\Delta\phi$ of 1.07° compared with 1.28° in the female. The facet size is also larger, as expected from diffraction considerations: 37 μm compared with 29 μm in the female. In houseflies and probably in other flies there are also anatomical differences at the receptor level that suggest that this region (the 'love spot' as it has been called) is specifically adapted for improved sensitivity. This is no doubt due to the very fast response times required for high-speed chasing. Male flies also have a number of 'male specific' interneurons in the optic ganglia, which are undoubtedly involved in the organization of pursuit behaviour.

In the small hoverfly *Syritta pipiens* the sex difference is particularly striking. In the male's acute zone $\Delta\phi$ is about 0.6°, nearly three times smaller than elsewhere in the eye, or anywhere in the female eye (Fig. 7.18a). Drone bees have a similar antero-dorsal acute zone, where the density of ommatidial axes is three to four times greater than anywhere in the female eye (Fig. 7.17b). They use this region when they chase the queen, and can be induced to chase a dummy queen on a string subtending only 0.32°, much smaller than the ommatidial acceptance angle of 1.2°. This implies that the trigger for pursuit is a brief decrease of about 6 per cent in the intensity received by single rhabdoms.

An increase in the detectability of small objects can be achieved either by reducing the rhabdom acceptance angle ($\Delta\rho$, Fig. 7.6) so that a small target causes a large change in the signal on the rhabdom that images it, or by increasing the numbers of photons available to the rhabdoms, thereby reducing the noise against which the signal must be detected. Either course of action requires a larger facet diameter D. In most of the examples discussed it seems that the increased facet diameter in the acute zone is 'spent' on reducing $\Delta\rho$, but in one well-documented case that is not so. The male blowfly *Chrysomyia megalocephala* has a 'bright zone' rather than an acute zone, where $\Delta\rho$ is similar to the rest of the eye, but the photon catch per rhabdomere is enhanced by an increase in both facet and rhabdomere diameter (van Hateren *et al.* 1989). This increase, compared with elsewhere in the eye, is about tenfold. It is not known why this fly has taken this particular route, but one would guess that it mates in dim conditions.

There is an interesting exception to the rule that it is always the males that have the acute zone. In pipunculid flies of the genus *Chalarus* the females have greatly enlarged ommatidia in the fronto-dorsal region. These flies parasitize leafhoppers (Homoptera) and the females have to locate these on the undersides of leaves in order to lay their eggs. The males have no equivalent need for keen eyesight.

Most of the animals just discussed have to detect their prey or mates against a background of foliage. This is a far from easy task, as the target usually only differs from the background by virtue of its motion, not because of static qualities such as brightness or texture. However, many insects and crustaceans have simplified the problem by using the sky as a background, against which any non-luminous object becomes a dark spot. Thus one finds not only upward-pointing acute zones, but also double eyes with one component directed skyward – or in the ocean, towards the surface.

Dragonflies hunt other insects on the wing, and have acute zones with a variety of configurations. Many in fact have two acute zones, one forward-pointing, and presumably concerned with forward flight as discussed above, and another directed dorsally and used to detect prey (Figs 7.17c and 7.18c). The migratory, fast-flying aeschnids have the largest eyes and most impressive acute zones. 28 672 ommatidia have been counted in one eye of *Anax junius*, which has the smallest inter-ommatidial angles of any insect (0.24° in the dorsal acute zone), and facets of corresponding size (62 μm). The dorsal acute zone takes the form of a narrow band of high resolution extending across the upper eye along a great circle, 50–60° up from the forward direction (Fig. 7.17c). The axis density (5 per square degree) is twice that in the forward acute zone, and five times higher than in a male blowfly. The dorsal acute zone is easily visible as a wedge of enlarged facets (Fig. 7.18c). Presumably the great high-acuity stripe in *Anax* is used to trawl through the air, picking out insects against the sky much as the scan line on a radar set picks up aircraft. 'Perching' dragonflies detect their prey from a stationary position on the ground or vegetation, rather than on the wing. In *Sympetrum* species there is a dorsally-directed acute zone with high acuity but low sensitivity, surrounded by a field of low resolution but high sensitivity. This seems to be an ideal combination for detecting fast-moving objects (with the sensitive periphery) and then fixating and tracking them with the acute central zone (Labhart and Nilsson 1995).

Simuliid flies have divided eyes, and use the upper part to detect potential mates against the sky. They can do this at a distance of 0.5 m, when a female subtends an angle of only 0.2°. As in drone bees, this is a small fraction of an acceptance angle. The eyes of male bibionid flies are similarly divided (Fig. 7.18b) with larger facets and smaller inter-ommatidial angles in the dorsal eye (1.6° compared with 3.7°, in *Bibio marci*). The upper eyes are used exclusively for the detection of females; movement of stripes around the lower eye evokes a strong optomotor turning response – the almost universal visual behaviour used by insects to prevent involuntary rotation – but the dorsal eye is quite unresponsive to this kind of stimulus (Zeil 1983).

Amongst the crustaceans, mid-water representatives of three groups have specialized in double eyes: the hyperiid amphipods with apposition eyes (Fig. 7.19), and the euphausiids and mysids with superposition eyes (see Chapter 8). The

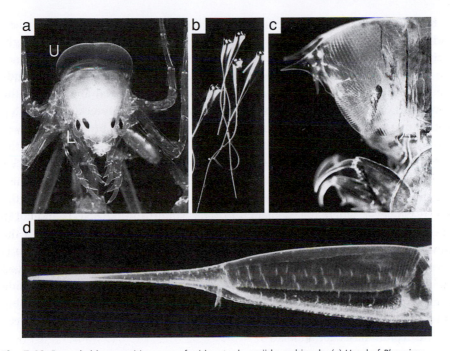

Fig. 7.19 Remarkable apposition eyes of mid-water hyperiid amphipods. (a) Head of *Phronima sedentaria*, from in front. The lenses of the upper eye cover the whole of the top of the head, and send light via 5 mm long light guides to the inner pair of dot-like retinae near the jaws. The outer pair of retinae serve the much smaller tear-drop-shaped lower eyes. (b) Lenses with attached light guides dissected from the upper eye of *Phronima* (see also Fig. 7.5d). (c) *Phrosina semi-lunata*, a relative of *Phronima*, also has a divided eye with separate retinae, but the halves are not separated. Total eye height 2.4 mm. Note the larger facets of the upper eye. The resolution distribution is shown in Fig. 7.17d. In both species the eyes are very transparent, and the retina is condensed to minimize its visibility. (d) Cylindrical eye of *Streetsia challengeri*. Like the other two species, the optical elements are transparent lens cylinders, but here arranged asymmetrically around the retina, which is the dark sausage-shaped structure in the ventral region of the eye. Curiously, the eye (*right*) has no field of view in the direction of the forward-pointing spike (*left*). The eye is 7 mm long, and the bearer of the eye off the page to the right is a surprisingly shrimp-like animal.

hyperiid eyes present an extrordinary range of eye anatomy, from surface-living forms with single eyes, to mid-water species with double eyes of various kinds (*Phrosina*, *Phronima*, and *Streetsia* are illustrated In Fig. 7.19), and finally to the deep-living *Cystisoma*, a large and very transparent animal which only has the upward-pointing component of the eye. The logic of this trend seems to be that the deeper an animal lives, the more important it becomes to devote as much photon-catching power as possible to the residual downwelling light, because it is against this dim background that potential food can be sighted from the silhouette it casts. It is interesting in this context that a great many mid-water animals disguise their silhouettes in various ways. Hyperiid amphipods do this by being transparent, whereas others such as the euphausiids (krill), many fishes, and some squid use ventrally directed photophores to substitute for the light blocked by the silhouette. In many cases the brightness of this bioluminescence can be adjusted to match the background light. In all the double-eyed hyperiids the upper part has larger facets and smaller inter-ommatidial angles than the lower part. *Phronima sedentaria*, a remarkable animal that protects itself with a transparent barrel hollowed out from the body of a salp, is probably the most extreme in this respect. The inter-ommatidial angle ($\Delta\phi$) is only 0.25° in the dorsal eye, compared with 10° in the ventral. Interestingly, the acceptance angles of ommatidia in the dorsal eye are about 9 times greater than the $\Delta\phi$ values, which in an ordinary terrestrial eye would imply huge over-sampling of the image. However, where the problem is to detect single dot-like objects and not to resolve texture, it can be shown that this is not a real mismatch, because the apparent contrast loss can be recovered by neural pooling later on. *Phronima* is also unique in that the light focused by the lenses of the dorsal eye is conveyed the 5 mm distance to the ventrally-situated retina by light-guides 18 μm wide (Figs 7.5d and 7.19b). The function of the lower eyes in *Phrosina* and probably other double-eyed hyperiids appears to be to detect and track luminous objects, such as bioluminescent animals.

Horizontal acute zones

As we have seen, many flying insects have a zone of increased vertical acuity around the horizon, no doubt reflecting the visual importance of this part of the surroundings. The visual field of the locust in Fig. 7.16 shows this clearly. There are environments where this region is even more important. Sand and mud flats are good examples, and many of the crabs that inhabit them have a narrow band of high vertical acuity around the equators of the eyes (Zeil *et al.* 1989). In the ghost crab, *Ocypode ceratophthalmus* (Fig. 7.10d) this band is about 30° high, with vertical inter-ommatidial angles as low as 0.5°; by contrast the horizontal inter-ommatidial angles are four times larger. There are interesting differences in this respect between crabs of the flat beach and those of the rocky upper shore. The former tend to have tall eyes close together, with a very pronounced equatorial band, whereas the latter have rounder eyes far apart, with a weakly developed band. Zeil *et al.* (1989) suggest that the tall-eyed crabs measure distance by the angle down from the horizon to the feet or base of the object they are looking at (a strategy which will only work on a flat surface) whereas the upper shore crabs with their wide-spread eyes use some form of binocular stereopsis. Another feature of the horizon is that objects that penetrate above it are necessarily larger than the crab itself, and are thus likely to be predators. Fiddler crabs (*Uca pugilator*) react defensively to moving objects above the horizon, but not to objects of similar angular size or speed below it (Layne 1998).

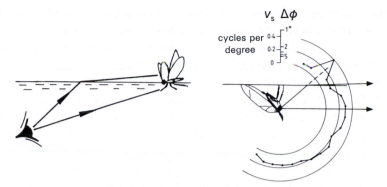

Fig. 7.20 Vision from below the surface film. *Left*: a floating object can be seen both above and below the surface. *Right*: distribution of resolution, expressed here as both inter-ommatidial angle ($\Delta\phi$) and acuity ($v_s = 1/2\Delta\phi$), around the eye of the back-swimmer *Notonecta*. The eye has two regions of elevated resolution, corresponding to the upper and lower faces of the water surface.

Insects that fly over water have a similarly narrow equatorial field of interest. Empid flies hunt close to the surfaces of ponds, again looking for stranded insects, and they have a horizontal acute zone that can be recognized by a linear region of enlarged facets around the eye (Fig. 7.18d). In *Rhamphomyia tephraea*, vertical inter-ommatidial angles are only 0.5° in this 15° high region, rising to 2° above and below it.

Water surfaces themselves provide a similarly constrained field of view, and water-striders (*Gerris*), which hunt prey stranded in the surface film, have a narrow acute band imaging this region, as shown in Fig. 7.16b. This has a height of only about 10°, centred on the horizon, and within this the vertical inter-ommatidial angle in the frontal region is only 0.55°, which is close to the diffraction limit, and impressive in an eye with only 920 ommatidia (Dahmen 1991). The backswimmer, *Notonecta*, is in some ways even more remarkable. Living just below the surface film, it looks up at the water surface, and can view potential prey in two ways. With the ventral part of the eye (*Notonecta* hangs upside down) it can look through the water surface to view the top of the prey in air, or it can look below the surface to see the bottom of the same prey through the water (Fig. 7.20). The two views are separated by about 30° in the sagittal plane. It turns out that there are actually two acute bands in *Notonecta* as there are in the fish *Aplocheilus* (Fig. 4.8), each imaging one of the views of the surface (Schwind 1980).

The anomalous eyes of Strepsipterans and Trilobites

We end this chapter with a discussion of a type of eye that seems to break all the rules of compound eye design, and which comes close to straddling the gap between simple and compound eyes. Strepsipterans are tiny parasites of wasps and other insects, in which the males take to the wing for a few hours in order to find mates. In *Xenos peckii* each eye of the male has about 50 lenses (compared with 700 in the similar sized *Drosophila melanogaster*, Fig. 7.21), but the lenses are large (65 μm diameter), and beneath each there is a 'retina' of about 100 receptors (Buschbeck *et al.* 1999). Thus each facet of this compound structure is quite unlike the ommatidia of other apposition compound eyes, and is actually a complete little eye (eyelet) in its own right, with a field of view of about 30°. Within each eyelet the inter-receptor angle is about 4°, which is comparable with other insects of similar size. A problem with an eye with this design is that the many inverted images do not join up. The problem is similar to that presented by the neural superposition eyes of Diptera (Fig. 7.4c), and the cure is the same: the images need to be re-inverted by an appropriate crossing over of the axons joining the receptors to the lamina – the first optic ganglion. Bushbeck *et al.* (1999) found that these crossings over (chiasms) do indeed exist, so the machinery is present for producing a single erect image, as in a conventional compound eye. There are no obvious ecological reasons why this odd group of insects should have evolved so strange an eye. Whilst it is tempting to think that each eyelet is resolving an image and that the sampling unit in this eye is a single rhabdom within the eyelet, the available evidence does not support this. Pix *et al.* (2000) made a comprehensive study of the optomotor response of these animals to moving patterns (see Fig. 4.2) and concluded that the sampling unit for this type of behaviour was the whole eyelet, not the individual receptor. If this is true for other behaviours (it may not be) then why should an eye with this design exist at all? The enigma continues.

It seems likely that eyes with this peculiar design have occurred once before, in the extinct phacopid trilobites, best known from the genus *Phacops* (Fig 7.21). Like *Xenos* their eyes have small numbers of large (0.5 mm) somewhat separated facets, and are known as 'schizochroal' eyes as opposed to the 'holochroal' eyes

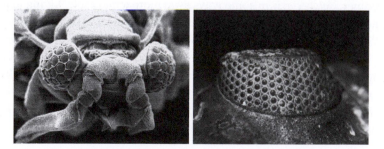

Fig. 7.21 Eyes of a strepsipteran insect (*Xenos peckii*) which has a small retina behind each facet (*left*), and a phacopid trilobite (*Hollardops mesocristata*) which probably had the same 'eyelet' design (*right*). The *Xenos* eye is about 0.3 mm in diameter (scanning electron micrograph by Elke Buschbeck and Birgit Ehmer); the *Hollardops* eye is 9 mm long (trilobite kindly identified by Pierre Morzadek). Notice the difference in size, number, and packing of the lenses compared with the conventional eyes shown in Fig. 7.18.

of the majority of trilobites (Levi-Setti 1993). The latter seem to be ordinary apposition eyes, but the huge lenses of the schizochroal eyes would only make sense as conventional compound eyes if the animals had lived in conditions of near darkness. The more likely explanation is that these were 'eyelet' eyes, like those of Strepsipterans.

Summary

1 Apposition compound eyes are made up of ommatidia, in which each receptor group receives an inverted image from its own lens. In conventional apposition eyes the receptive rod (rhabdom) in each ommatidium does not resolve detail within each image, but acts as a detector that measures the average brightness of a small region of space, typically about 1° across. The overall erect image seen by the animal is the mosaic formed by these adjacent fields of view.

2 In dipteran flies the situation is slightly different: the inverted image in each ommatidium *is* resolved by seven separate receptors. However, the responses of these are combined in the lamina (first synaptic layer) in a way that pools their signals, giving enhanced sensitivity without loss of resolution. As far as the fly is concerned the form and resolution of the overall image is the same as in a conventional apposition eye. This arrangement has been called 'neural superposition'.

3 Because individual facet lenses are very small the images they produce are severely limited by diffraction, so that the minimum resolvable angle is rarely better than 1°. To improve on this requires larger lenses as well as more of them, and the size of the eye rapidly becomes unsupportable. Arthropods do achieve enhanced resolution, however, by having local regions of enlarged facets and closer ommatidial axes, at the expense of resolution elsewhere.

4 Much can be learnt about the way that apposition eyes sample the surroundings from a study of the pseudopupil: this is the small dark spot that appears to move across the eye as the observer moves around it.

5 Acute zones are found frontally in many flying insects; dorsally or dorsofrontally in insects that capture other insects on the wing, either to mate with them or to eat them; and around the horizon in arthropods that live in a flat environment, such as crabs on a beach, or bugs that hunt in the surface film of ponds.

8 | Superposition eyes

Introduction – the nature of superposition imagery

From the outside, apposition and superposition eyes are almost indistinguishable. Both are convex structures with facets of similar dimensions, and are clearly variants of the same general design. But there the resemblance ends. Internally there are several crucial anatomical differences: the retina is a single sheet, not broken up into discrete ommatidial units as in apposition eyes, and it lies deep in the eye, typically about halfway between the centre of curvature and the cornea. Between the retina and the optical structures beneath the cornea there is a zone with very little in it, the *clear zone*, across which rays are focused – the equivalent of the vitreous space in a camera-type eye (Fig. 8.1). The optical devices them-

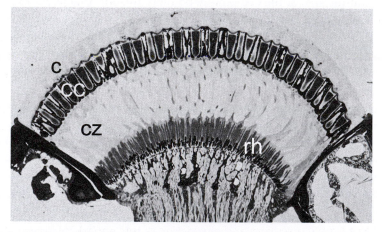

Fig. 8.1 Section through the refracting superposition eye of a nocturnal dung beetle, *Onitis westermanni*, showing the cornea (c), the outer row of crystalline cones (cc), the wide optically unencumbered clear zone (cz), and the convex rhabdom layer (rh) about halfway out from the eye's centre of curvature. Photograph by Dr S. Caveney.

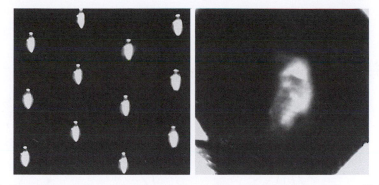

Fig. 8.2 Apposition and superposition images. The photograph on the left shows the multiple inverted images of a candle flame, taken through the facets of the eye of a robber fly. The erect image on the right, of an influential nineteenth-century naturalist, was taken through the cleaned cornea of the eye of a firefly, *Photuris* sp. Both were taken with the cornea in air, and the region behind the optics in physiological saline.

selves are various; as we shall see, they may be refracting telescopes, mirrors, or lens–mirror combinations, although to a cursory examination most do not look very different from the lens structures of apposition eyes.

The real surprise is optical. All superposition eyes produce a single deep-lying *erect* image in the vicinity of the retina. Not only does this distinguish them from apposition eyes, which have multiple inverted images, but also from camera-type eyes where the image is inverted. Clearly we are dealing here with something quite out of the ordinary. Around the turn of the twentieth century there were a number of successful attempts to photograph these images. There is one in Exner's monograph of 1891, and a delightful portrait taken by H.E. Eltringham of his friend Sir Edward Poulton 'taken through the eye of a glow-worm' and reproduced in Imms' *Insect natural history* (1956, Plate VIIb)). Our recent attempt to recreate this photographic feat, in a firefly eye, is shown in Fig. 8.2 (right), where the single erect image is contrasted with the multiple inverted images of an eye of the apposition type. It turns out that it is important to use a beetle (such as a firefly) for this. Other insects, in particular moths, have superposition eyes, as do crustaceans such as krill (euphausiids), but there the optical structures that create the image are not joined to the cornea, and they are swept away when the eye is cleaned to make a lens for photography. In beetles, however, the optical elements are continuous with the cornea and so survive the removal of the eye's internal structures.

The credit for the discovery and elucidation of this remarkable piece of optics is due to Sigmund Exner, who worked on the problem throughout the 1880s and published his complete findings in 1891. Exner showed that the only way an erect image could be formed was for the optical elements to behave in a rather

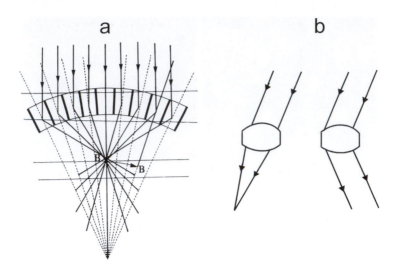

Fig. 8.3 (a) Exner's diagram of 1891 showing the ray paths needed to produce an image (B–B) in a superposition eye. Note the 'dog-leg' way in which the rays must be bent by each optical element. (b) ray bending by a conventional lens, and by an element in a superposition eye. Ordinary lenses cannot perform the required task.

strange way, as shown in Fig. 8.3a. Basically what each has to do is not to form an image from a parallel beam as in a conventional lens, but to redirect light back across the element's axis, to form another parallel beam on the same side of the axis (Fig. 8.3b). Exner realized that although a single lens wouldn't do the job, a two-lens telescope would, and he went on to demonstrate (as well as he could with the technology of the time) that such structures were indeed present in the superposition eyes of insects. In the 1950s and 1960s Exner's ideas ran into difficulties, when, armed with a new device called an interference microscope, scientists started to look for high refractive index structures that would make Exner's telescopes a reality. Unluckily, some of the first studies were on crayfish eyes (certainly superposition eyes by their clear-zone anatomy) but with nothing that could serve as a lens or lens combination. What should have been the optical elements appeared to be squarish blobs of low-refractive index jelly, with no promising optical properties (see Fig. 8.13c). A plethora of unsatisfactory theories arose as to how these eyes might work, but as it turned out they were not necessary. In 1975 Klaus Vogt discovered that crayfish and their relatives use a system that works with mirrors, not lenses. However, for all the other eyes to which Exner had addressed his attentions the refracting telescope mechanism was, and still is, the correct one. Later still, a third mechanism was discovered in certain crabs that combined both refracting and reflecting elements (Nilsson 1988). These three ways of achieving the 'dog-leg' ray-paths required for superposition imagery (refracting, reflecting, and mixed or parabolic

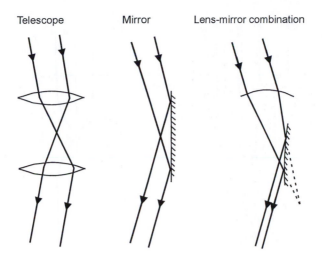

Fig. 8.4 Three arrangements capable of 'dog-leg' ray bending (see Fig. 8.3). A two-lens telescope (*left*), a plane mirror (*centre*), and a combination of lens and curved mirror (*right*). These correspond to the elements in the three known types of superposition eye: refracting, reflecting, and parabolic superposition.

superposition) are shown in Fig. 8.4. In the sections that follow they are discussed in turn.

Refracting superposition

Telescopes and lens cylinders

In a lens-based superposition eye the optical elements need to act as simple inverting telescopes which redirect the entering beam of light back across the axis, as shown in Fig. 8.3b. The most straightforward way to do this is to have two lenses separated by the sum of their individual focal lengths, with an image plane between them (Fig. 8.4). Exner realized that, given plausible refractive indices and the curvatures of the structures revealed by histology, there was not enough ray-bending power in each element of a beetle eye to make this possible. He came up with an idea which was similar to Matthiessen's solution for the fish lens (Chapter 4), namely, that the structures must have an internal refractive index gradient. The result would be that most of the ray bending would occur within the tissue, rather than at its external surfaces. The pure form of this structure, a flat-ended cylinder with a radial parabolic refractive index gradient, Exner called a *lens cylinder*. He showed that, depending on its length, it could act as a single lens (Fig. 8.5a) as in the apposition eye of *Limulus* (Chapter 7), or as a pair of lenses making up an inverting telescope of the kind required for superposition optics (Fig. 8.5b). Although Exner did not have the means in his time of establish-

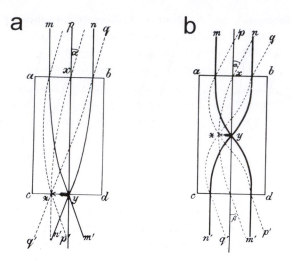

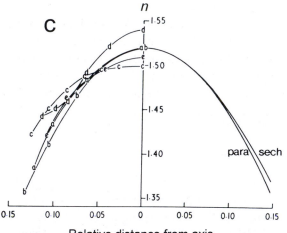

Relative distance from axis

Fig. 8.5 Lens cylinders. (a) and (b). Exner's diagrams of lens cylinders capable of producing a simple inverted image, as in the apposition eye of *Limulus*, and dog-leg ray redirection as in the superposition eyes of moths and beetles. Essentially, the single lens structure in (a) turns into (b) if its length is doubled so that rays producing the first image are brought parallel again (i.e. re-collimated). Note that (b) is analogous to the two-lens telescope in Fig. 8.4. (c) *Right*: two versions of the refractive index gradient (*n*) from centre to periphery of a lens cylinder, required to produce the kind of imaging shown in (a) and (b). The abscissa is the radial distance from the axis divided by twice the focal length. The parabolic and hyperbolic secant gradients differ only slightly in their optical properties. *Left*: recent measurements made by interference microscopy of the refractive index gradients in a variety of lens cylinder structures in compound eyes. (*a*, euphausiid; *b*, firefly; *c*, moth; *d*, skipper butterfly; *e*, *Limulus*). The measured and theoretical estimates of the gradient agree very well.

ing whether beetles and moths had optical elements with the required refractive index gradient, numerous studies since the advent of interference microscopy have shown that his brilliant conjecture was correct (Kunze 1979; Nilsson 1989). Figure 8.5c gives a selection of these measurements. Interestingly lens cylinder structures were invented *de novo* in the 1970s, manufactured from both glass and plastic by processes that provided axial refractive index gradients. They are now used in optical fibre coupling devices, and the lenses of some CD players.

Resolution and sensitivity

The geometrical optics of a superposition eye are shown in Fig. 8.6. The peculiarities of this type of image formation mean that the nodal point of the eye (the point through which rays pass undeviated) is at the centre of curvature, and the focal length is the distance out from the centre to the image. This conforms to the general definition of focal length (f) given in eqn (3.1):

$$O/U = \alpha = I/f$$

where O and I are object and image sizes, U is the (large) object distance, and α is the angle in radians subtended by object or image at the nodal point. The inter-rhabdom angle ($\Delta\phi$) is s/f, where s is the rhabdom separation, just as in a camera-type eye. As in apposition eyes, the rhabdom acceptance angle is a combination of the geometrical subtense of a rhabdom (d/f), and the width of the blur circle provided by the optics (see the discussion in Chapter 7, and Fig. 7.6).

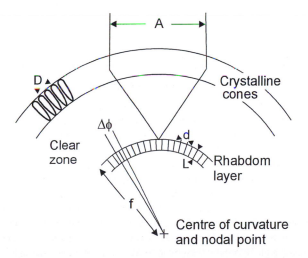

Fig. 8.6 Optical definitions that apply to superposition eyes. A, diameter of the superposition pupil. D, facet diameter (note that A replaces D in the sensitivity eqn 3.6); f, focal length; d, rhabdom diameter; $\Delta\phi$, inter-receptor angle ≈ d/f; L, rhabdom length.

In the past there has been a belief that superposition eyes suffer from poor resolution, mainly because of the difficulty of conceiving how the large numbers of ray bundles contributing to a single point on the image could be directed to the receptor layer with sufficient accuracy. However, this reputation seems not to be justified, except perhaps in extreme cases. In a careful study of the eyes of dung-beetles that fly at different times of the day and night, McIntyre and Caveney (1998) found that in the day-flying *Onitis belial* about 50 optical elements (the effective superposition aperture) contributed to the image at any one point, and in the nocturnal *O. aygulus* the number was close to 300. *O. belial* had a calculated rhabdom acceptance angle ($\Delta\rho$) of 2.2°, which is comparable with values from many apposition eyes, and in *O. aygulus* $\Delta\rho$ was somewhat larger, 3.0°, which is still quite impressive for an eye with such a huge aperture. These modelling studies have since been confirmed by electrophysiological recordings from single receptors. In the Australian day-flying moth *Phalanoides tristifica* the image quality has been measured directly with an ophthalmoscopic method which uses the eye's own optics to view the retina and images on it (Fig. 8.7). The result was that $\Delta\rho$, the acceptance angle of a rhabdom when viewing a point in space, was 1.58°, of which the optical point-spread function contributed only 1.28°. This is itself only slightly larger than the half-width of the Airy diffraction image from a single facet. Thus a superposition eye in which 140 elements contribute to a point image has optics that are almost as good as is physically possible. Other day-flying moths (including skipper butterflies) show similar excellent image quality. (Although the superposition pupil is many times wider than an individual facet, this does not decrease the size of the Airy diffraction image, as it

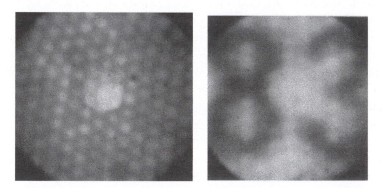

Fig. 8.7 Image quality in a refracting superposition eye. *Left*: the rhabdom mosaic of the eye of the day-flying moth *Phalanoides tristifica*, viewed through the eye's own optics using an ophthalmoscope (Land 1984). Each rhabdom is enclosed in a reflecting sheath and so appears light compared to the brown surrounding pigment. The inter-receptor angle ($\Delta\phi$) is 1.9° and the physical separation of the rhabdoms is 16 μm. *Right*: image on the moth's retina of an object (the year the photograph was taken) in the moth's field of view. This gives a good impression of the image quality in a superposition eye. The figures are 17° high.

Table 8.1 Sensitivity calculation for apposition and superposition eyes of the same size and resolution

Parameter	Apposition	Superposition
radius (r)	1 mm	1 mm
inter-receptor, or interommatidial angle ($\Delta\phi$)	2° 0.035 rad	2° 0.035 rad
focal length (f)	0.1 mm	0.5 mm
receptor separation (superposition: $s = f\Delta\phi$)	–	17.5 μm
receptor diameter (superposition: $d = s$) (apposition: $s = f\Delta\rho$, where $\Delta\rho = \Delta\phi$)	3.5 μm	17.5 μm
Aperture (apposition $D = r\Delta\phi$) (superposition: $A = 10$ facet diameters)	35 μm —	— 350 μm
Sensitivity (eqn 3.6) $S = 0.62\ D^2\Delta\rho^2$ or $0.62\ A^2\Delta\rho^2$	0.93 (μm^2)	93 (μm^2)

would if the aperture of a lens eye were increased. This is because the ray bundles from different facets travel different optical distances to the image (unlike the rays in a single large lens) and so do not interfere constructively at the image point. The Airy disc diameter thus depends on the diameter of single facets, just as in apposition eyes.)

Size for size, superposition eyes are more sensitive than apposition eyes, which is why they are most commonly encountered in animals such as moths and fireflies that are active at night, or in marine crustaceans from the mid-water depths where the light regime is similar to moonlight on the surface. To quantify the sensitivity difference we should consider eyes of similar size, and the same resolution (the same $\Delta\phi$). The calculation is given in Table 8.1. The additional assumptions are made that the effective pupil in the superposition eye is 10 facets wide, and that the focal length of the apposition ommatidium is 0.1 mm. These are both realistic values. The result is that the superposition eye is a hundred times more sensitive than a similar sized apposition eye, and in truly nocturnal moths and beetles, which have even larger superposition pupils, the sensitivity can be ten times higher again. Because rays entering the outer parts of the superposition pupil are less effective than central rays, these figures somewhat overestimate the sensitivity gain of superposition eyes, but not by more than a factor of two.

Fig. 8.8 Many superposition eyes show eye-glow when observed from the same direction as the illuminating beam. Parallel light is focused to a spot on the retina, and then reflected back by the tapetum to emerge through the same super-position pupil that it originally entered. *Left*, dark adapted reflecting superposition eye of a decapod shrimp (*Leander*); *right*, light adapted eye in which rays from the outer zones of the pupil have been cut off (Fig. 8.9) leaving a small dark central pseudopupil.

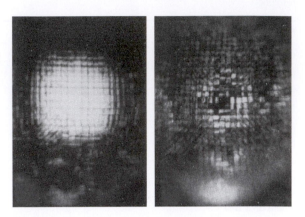

Eye-glow and the superposition pupil

Most moths, and some decapod crustaceans with reflecting superposition eyes, have a reflecting layer (tapetum) behind the rhabdoms. Its function is the same as the tapetum in the eye of a cat: to double the light path through the photo-receptors and so improve their photon catch. In some diurnal moths a reflector also surrounds each rhabdom, optically isolating it from its neighbours. In dark-adapted eyes the tapetum causes the eye to glow when viewed from the same direction as the illuminating beam. In some diurnal moths the glow is always visible (*Macroglossum*, Plate 3). The mechanism is similar to that in a cat's eye. The optical system forms a point image of the light source on the tapetum, or close to it, and this point acts as an emitter of light which, on passing through the optics again, emerges as a roughly parallel beam.

If the optics are good, that is to say they really do bring a parallel beam to a point in the image, then the patch of glow seen at the surface of the eye will have the same diameter as the beam that entered the eye. This is the superposition pupil – the amount of eye surface from which rays contribute to each point on the image (Fig. 8.8). Eye-glow can also provide a useful test of image quality. If the glow can only be seen over a narrow angle (a few degrees) from the direction of the illuminating beam, then the retinal image must itself be very small. On the other hand, if the glow can be seen over a wide angle (as is the case with many deep-sea shrimps, for example), this indicates either that there is a large blur circle on the retina, or that the tapetum is situated a long way from the focus.

Light and dark adaptation

The high sensitivity of most superposition eyes means that they must protect their visual pigment in daylight, and so need adaptation mechanisms that can

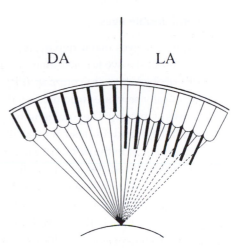

DA   LA

Fig. 8.9 Dark adaptation in a superposition eye. The most common dark adaptation mechanism involves the inward migration of dark pigment from between the crystalline cones. This progressively cuts off more oblique rays from the outer zones of the superposition pupil and so reduces the light flux on the retina.

reduce image brightness by several orders of magnitude. The main mechanism of light adaptation in superposition eyes consists of pigment movements that result in the progressive interception of rays from the outer zones of the superposition pupil (Fig. 8.9). This reduction may ultimately result in light from only a single facet reaching a single point in the image, which is essentially the apposition condition.

The eye-glow (Fig. 8.8) provides a means of monitoring the process of light and dark adaptation. As oblique rays across the clear-zone are cut off during light adaptation (Fig. 8.9), so the brilliance of the glow and the size of the patch reduce, often disappearing completely. In the dark they slowly return.

In insects with refracting superposition eyes the main pigment movement is a longitudinal inward migration of granules in both the primary and secondary pigment cells (see Autrum 1981). In the dark the granules are bunched up between the crystalline cones, and with the onset of light they extend inwards, over a matter of minutes, to occupy much of the clear zone. In many crustaceans, especially decapods such as crayfish with reflecting superposition eyes, there is also an outward movement of pigment in the proximal pigment cells. In the dark the pigment is held beneath the basement membrane, but in the light it moves up between the rhabdoms, preventing rays from entering them obliquely and thus reducing the width of the cone of light that each rhabdom can accept.

Interestingly, the trigger for pigment migration in some moths is not provided by photoreception in the rhabdoms themselves. In the crepuscular sphingid moth *Deilephila*, Nilsson *et al.* (1992) showed that a region immediately beneath each crystalline cone initiates pigment migration, when illuminated with ultraviolet light, and that the much deeper-lying rhabdoms are not involved. However, in the owl-fly *Ascalaphus*, a day-flying neuropteran with double superposition eyes, the pigment movements can be triggered from both the region below the cones, and also from the rhabdoms themselves.

Single and double eyes

In superposition eyes major departures from spherical symmetry are rare because the geometry of the eye is constrained by the shared optics. However, in euphausiids (krill) bilobed eyes are common (Fig. 8.10 a and b). The two components are usually optically separate structures, so that each can be regarded as a separate eye with its own symmetry. The focal length of the dorsal eye is usually longer than that of the ventral eye, and this results in smaller inter-ommatidial angles: in *Stylocheiron maximum* $\Delta\phi$ is 1.2° in the dorsal eye compared with 2.6° in the ventral. In *N. boopis* the lower eye is almost absent – a parallel with the amphipod *Cystisoma* (Chapter 7). It has been shown in the double-eyed *Nematoscelis atlantica* that the dorsal eye is always kept pointing upwards, towards the daylight, while the animals themselves swim at various angles to the vertical (Land 1980). Thus it seems very likely that the role of the dorsal eyes is to look for dark silhouettes against the downwelling light, as suggested for the hyperiid amphipods. The function of the wide angle, low resolution ventral eye is less clear, but given that it images the dark of the abyss its most likely role is the detection of bioluminescent objects.

In the genus *Stylocheiron* some species show a reduction in the numbers of facets in the dorsal eye, and this seems to be related to the depths at which the animals swim. In deeper-living species there are typically several hundred facets altogether. In *S. elongatum*, which lives at 180–420 m in daytime, each row contains 13–16 facets, but in *S. affine* (40–140 m) this reduces to 4–8, and in *S. suhmi* (0–50 m) there are only 3. This appears to be a crude but effective way of ensur-

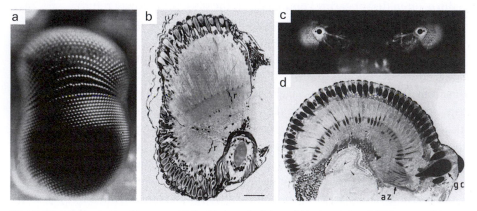

Fig. 8.10 Double superposition eyes. (a) Double eye of a living mid-water euphausiid *Nematoscelis megalops*, eye height 2.5 mm. (b) Section through the eye of *Nematoscelis atlantica*, height 0.9 mm. Note the two separate retinae (*r*) and the much larger clear zone in the upper eye. The structure at the bottom right is a photophore, whose function is to disguise the eye's silhouette by counter illumination. (c) The eyes of the mysid *Dioptromysis paucispinosa* seen from behind, showing the single giant crystalline cone (44 μm) surrounded by the (16 μm) crystalline cones of the conventional eye. (d) Horizontal section of the eye in (c) showing the ordinary superposition eye (compare with Fig 8.1) and the giant crystalline cone (*gc*), with its own higher-resolution retina, or acute zone (*az*). From Nilsson and Modlin (1994).

ing that the different eyes provide similar retinal illumination levels. The ultimate reduction of the superposition eye design, to only one facet, is seen in a
tropical shallow-water mysid *Dioptromysis paucispinosa* (Fig. 8.10c). Nilsson and
Modlin (1994) describe this shrimp as 'carrying binoculars', which is an almost
exact analogy. The eyes are double: the main part is a conventional superposition
eye, but the accessory region is quite different. It has only one large facet with a
single giant crystalline cone, beneath which is a retina of 120 rhabdoms, compared with 800–900 in the rest of the eye. This accessory eye is thus a unique
example of a single-lens superposition eye – effectively a simple eye but with
erecting rather than inverting optics. The accessory eye functions as an acute
zone, with a minimum separation of rhabdom axes ($\Delta\phi$) of 0.64° – impressive in
an eye so small. The 44 μm diameter of the giant facet means that the diffraction
limit is much lower than in the rest of the eye, which has 16 μm facets. The most
peculiar feature of these already strange eyes is that the giant facet and its acute
zone normally point backwards! Nilsson and Modlin (1994) found that occasionally the eyes are rotated, directing the acute zones forwards, where they are presumably used for a higher resolution scrutiny of potential food or mates. Unlike
other double-eyed arthropods, there are no good reasons for thinking that these
special eyes normally point upwards.

Double superposition eyes are uncommon amongst insects. As mentioned
earlier, owl-flies (*Ascalaphus*) have double superposition eyes. Male mayflies have
a pair of dorsal superposition eyes (Plate 3), which they use for sighting females
against the sky, in a similar way to bibionid flies (Fig. 7.18b). However, the
lower eyes, present in both sexes and responsible for other visual activities,
are of the apposition type. Like the euphausiid eyes the field of view of the dorsal
eye is small, and it is adjusted to the environmental circumstances of the species:
those swarming in woods with small gaps in the canopy have the narrowest
fields.

The hummingbird hawkmoth, *Macroglossum stellatarum*, a spectacular fast-
flying diurnal nectar feeder, does seem to have overcome the spherical constraints of classical superposition optics (Warrant *et al.* 1999). It has a visibly
non-spherical eye across which there are considerable variations in resolution
(Plate 3). These are not reflected in the pattern of facet sizes, as is often the case in
apposition eyes (Chapter 7) but in the spacing of the retinal receptors, and also in
variations in focal length across the eye. The effect is to produce an anterior acute
zone coupled with a horizon streak, which is very similar to the pattern of resolution across the eye of a butterfly (Fig. 7.14). How the optical variation is achieved
without compromising image quality (which is excellent throughout the eye) is
not yet known, but presumably this entails a systematic variation across the eye
of the angular magnifications of the crystalline cones themselves. So far this is the
only known superposition eye that does depart substantially from a spherical
shape, without actually becoming divided.

Superposition and afocal apposition: the eyes of butterflies

Butterflies and moths are classified together in the Lepidoptera, and are undoubtedly very closely related. Most butterflies (skippers (Hesperidae) are the exception) have eyes that behave in most respects as apposition eyes. They have long narrow rhabdoms abutting the bases of the crystalline cones, no clear zone, and complex pseudopupils (Fig. 7.10). Many moths, on the other hand, have refracting superposition eyes with wide, deep-lying rhabdoms, clear zones and eye-glow. Transitions between the eye types must have occurred a number of times within the moths, as well as between moths and butterflies. A very similar picture emerges in the beetles, most of which have apposition eyes, but a substantial number of nocturnal and crepuscular groups, including the dung beetles and the fireflies, have superposition optics.

It is not very easy to see how it is possible to get from one type of eye to the other, without going through an intermediate which doesn't work. Apposition

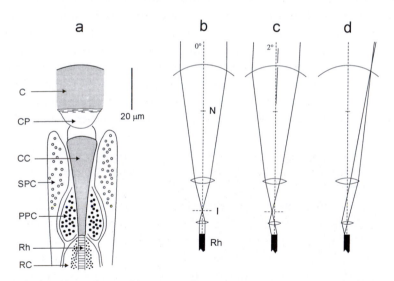

Fig. 8.11 Optics of butterfly eyes. (a) Anatomy of a single ommatidium of a butterfly. C, cornea; CP, corneal process; CC crystalline cone; SPC, secondary pigment cell; PPC, primary pigment cell; Rh, rhabdom; RC, receptor cell (b). The ommatidial optics are here represented by three lenses: the cornea using surface refraction (nodal point at N), a weak lens representing the distal part of the crystalline cone, and a strong lens in the proximal stalk (see Fig. 8.12b). Their combined effect is to produce a system with an internal focus (I) and a parallel output beam that matches the diameter of the rhabdom (Rh). (c) The same system with an input beam off axis by 2°. The resulting output beam is 12.8° off axis, close to the limit that the rhabdom can accept by internal reflection. (d) The optical system has the secondary property that the rhabdom tip is imaged on the cornea. This explains why waveguide modes produced by interference in the rhabdom can be seen, magnified, when the cornea is viewed by light reflected back through the rhabdom by the mirror at its base (Fig. 8.12c and Plate 3).

eyes use simple lenses and superposition eyes two-lens telescopes (or the equivalent lens cylinder devices), and there does not seem much room for compromise.
In the case of butterflies we do know the answer: their apposition eyes actually
have an extreme form of superposition optics in the ommatidia, in which the
proximal lens in each telescopic pair has become not weaker, as one might have
guessed, but extremely powerful (Nilsson *et al.* 1988).

The way this works is shown in Fig. 8.11. As in a normal superposition eye a
combination of the curved cornea and a weak lens cylinder in the distal region of
each crystalline cone results in the formation of an image within the crystalline
cone, about 10 μm in front of its proximal tip. This focused light then encounters
a lens with an extraordinarily short focal length – about 5 μm. This lens has thus
a power ($1/f$, where f is in metres) of 200 000 dioptres, or 10^5 times the power of a
pair of reading glasses. The discovery of this lens involved taking thin frozen
sections from the tiny region at the base of the crystalline cone, and examining
their image-forming properties. Figure 8.12a, where a crystalline cone is seen
against a wing scale, gives some idea of the size of the structures involved. To
our delight the last 10 μm of the cone produced excellent images (Fig. 8.12b),
from whose size we could work out the optical power (Nilsson *et al.* 1988). The
effect of this second lens is to bring the light focused by the first (distal) lens back
into a parallel beam (Fig. 8.11 b and c), again just as in a superposition eye. The
essential difference is that, whereas in a superposition eye the magnification of
the telescopic pair of lenses rarely exceeds –2, here it is much greater. The large
difference in the focal length of the distal and proximal lenses gives an overall
magnification of –6.4, in the nymphalid butterfly *Heteronympha merope*.

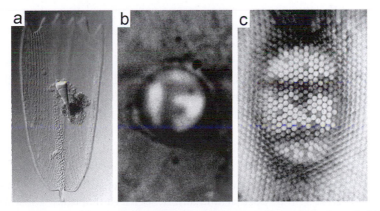

Fig. 8.12 Butterfly eyes. (a) A single crystalline cone from the eye of a small blue butterfly (*Zizina
labradus*) seen against a wing scale. Cone length is 40 μm. (b) The image of a letter F produced by a
5-μm-thick slice from the proximal part of a crystalline cone from the butterfly *Heteronympha
merope*. At this point the cone is only 5 μm wide. (c) The images of 2 lines, 10° apart, seen in the
corneal facets of *Heteronympha*. The images result from light that has entered the rhabdoms, been
reflected from the mirrors at the base of each rhabdom, and re-emerged from the rhabdom tip.

This high magnification has two important consequences, illustrated in Fig. 8.11. The first is that the beam that emerges from the proximal tip makes an angle with the axis that is 6.4 times greater than the beam that entered the facet from outside. A ray making an angle of 1° with the facet axis emerges at 6.4°, and similarly a beam 3° wide at the cornea (inversion) wide emerges into the rhabdom as a 19.2° wide beam. The significance of this is that a rhabdom with a refractive index of 1.36 will just contain (by total internal reflection) a beam 22° wide, which in turn means that the acceptance angle of the ommatidium will be limited to just over 3°: light making higher angles with the rhabdom wall will escape and be absorbed by the surrounding pigment. Thus in this kind of eye the ommatidial acceptance angle is limited principally by the refractive index of the rhabdom, not (as in a conventional apposition eye) by its diameter (Fig. 7.6). The second effect of the magnification is to reduce the diameter of the beam leaving the base of the crystalline cone by a factor of about 9 (angular magnification × refractive index), compared with that entering the facet. The entering beam is limited by the facet diameter, typically about 20 μm. The beam leaving the crystalline cone and entering the rhabdom is squashed down to a diameter of 2.1 μm, which is indeed close to the diameter of a butterfly rhabdom. Thus rhabdom diameter and facet diameter are related, and between them determine the effective aperture of the ommatidium, and hence its sensitivity. Bright-light butterflies tend to have smaller facets (20 μm) and narrow rhabdoms (1.5–2 μm), whereas the crepuscular Australian butterfly *Melanitis leda* has 35 μm facets and 5 μm rhabdoms.

Sadly, the story is yet more complicated. The small dimensions of the rhabdoms mean that diffraction and waveguide mode effects slightly alter the geometric story of the last few paragraphs. As this kind of apposition eye does things almost the opposite way round from a conventional apposition eye, none of the discussion in Chapter 7 is strictly appropriate. The effects of these wave-optic phenomena are discussed in some detail elsewhere (Nilsson *et al.* 1988) and here we will only mention one, the appearance of waveguide modes at the cornea. As Fig. 8.11d shows, and as we have seen in the last paragraph, the facet lens is imaged onto the rhabdom tip, reduced by a factor of 9. Light paths are reversible, so it is also true that the rhabdom tip is imaged, magnified, in the plane of the facet lens. Thus if there are interesting optical phenomena in the region of the rhabdom tip we should see a magnified version of them at the cornea. This turns out to be true in a rather spectacular way (Plate 3). In Chapter 3 there was some discussion of waveguide modes, the patterns that result from the interference of light trapped within a fibre such as a rhabdom. These are seen beautifully in the facets of butterflies. The other feature of butterfly eyes that makes this possible is the mirror-like tapetum at the base of each rhabdom (Fig. 6.10d). A narrow beam of light directed down the axis of a facet is guided through the rhabdom to the mirror, is reflected back up the rhabdom, and the small proportion that has not been absorbed emerges from the tip. This beam, with its mode structure, is displayed in the facet. The appearance of the mode patterns depends on the butterfly. All have the simplest (first-order) pattern which is bright in the middle and dimmer towards the edges. Larger butterflies (most of the nymphalids) with wider rhabdoms also show the more complex bi-lobed second-order mode (shown in Plate 3), and the crepuscular *Melanitis*, with the widest rhabdoms, has higher-order modes that give a pattern that is almost uninterpretable.

Interestingly, these mode patterns change as the eyes dark and light adapt. As in dipteran flies (Chapter 7), butterflies have dark pigment in the region around the rhabdom, and this moves into contact with the rhabdom wall at high light levels (Fig. 7.9b). This absorbs the portion of the modal light which travels outside the rhabdom (Fig. 3.7), and there is more of this extra-rhabdom light in the higher-order modes. These then disappear, leaving only the first-order mode. As the higher-order modes are wider (in angular spread) than the first-order mode, their loss has the beneficial effect of reducing the acceptance angle of the ommatidium in bright light, thus improving acuity. In *Melanitis*, with the widest rhabdoms, the effect is a halving of the acceptance angle from 3° to 1.5° (Land and Osorio 1990).

Another consequence of the presence of a mirror at the base of each rhabdom is that the apposition image can actually be seen at the eye surface (Fig. 8.12c). This is because the pattern displayed across the facets is an attenuated version of the light that has entered each rhabdom, traversed it twice and re-emerged from its distal tip. This image can only be seen if the eye is illuminated from a wide source, and it fades in a few seconds as the pupil mechanism bleeds light out of the rhabdoms.

What we have seen is that butterfly eyes behave as apposition eyes, because light entering a single facet is received by a single rhabdom. They are called 'afocal' because light is not focused on the rhabdom tip as in most apposition eyes, but enters the rhabdom as a parallel beam. In their fundamental optical design, however, these ommatidia remain of the superposition type, constructed from two-lens telescopes. This makes it easy to understand how different lepidopteran groups managed to switch readily from the diurnal (apposition) version of the afocal eye to the nocturnal (superposition) version. To become nocturnal, the powers of the distal and proximal lenses must become more equal, the receptor layer moves to a deeper location, and gradually more and more facets contribute to the image. There are no blind intermediaries.

There still remains a problem of origins. By common consent, the first compound eyes in the wormy ancestors of the arthropods had to be of the orthodox, focal, apposition, type (Nilsson 1989). The butterfly apposition eye helps us to understand the relationship between apposition and superposition optical types, but where did *it* come from? Could it have originated from an ordinary apposition eye by a gradual increase in the refractive index of the proximal part of the crystalline cone? This need have had no deleterious consequences to image formation; indeed the 'afocal' type of apposition eye can resolve marginally better than the ordinary 'focal' type. Or it may be that butterflies developed their unique type of eye from the superposition moth eye, which had already acquired superposition optics by some other route. The problem remains unsolved.

Reflecting superposition

There was a period, between about 1955 and 1975, when shrimps and their relatives couldn't see. The use of interference microscopy in the 1950s had shown that the optical structures that should have been producing the images in these eyes had none of the required qualifications. Instead of being lens cylinders with high refractive indices and a radial gradient, they were square structures of low refractive index, made of more or less homogeneous jelly (Fig. 8.13c). This is hardly a good basis for any kind of optical system. The solution to this enigma was first provided by Klaus Vogt in 1975, working on crayfish eyes (a full account is provided in Vogt 1980). He found that the jelly blobs were silvered, and that they were not lenses at all, but mirror boxes (Fig. 8.13b). Shortly afterwards the same mechanism was found in a shrimp, and it now appears that this reflecting system is the rule throughout the long-bodies decapod crustaceans – the shrimps, prawns, lobsters, crayfish, and the anomuran squat lobsters. The hermit crabs and the true crabs (Brachyura), however, have either apposition or parabolic superposition eyes (see below). The reflecting mechanism does not occur outside the Decapoda; even the euphausiids, sister group to the decapods, have refracting superposition eyes that resemble the eyes of moths much more closely than those of decapod shrimps.

In essence the reflecting superposition mechanism is extremely simple. In 1975 Vogt wrote:

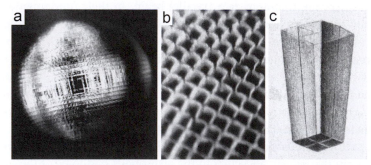

Fig. 8.13 Reflecting superposition eyes. (a) Eye of the decapod shrimp *Palaemonetes varians*. Note the square facet array, the silvery appearance, and the dark central facets of the region contributing to the image in the light adapted state. (b) Distal tips of the mirror boxes in the eye of a living crayfish. (c) Tapered mirror box in a shrimp (*Palaemon squilla*) drawn by Grenacher in 1879. The structure is 63 μm deep and 30 μm along each top edge.

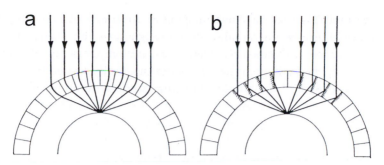

Fig. 8.14 Comparison of ray paths in a refracting (a) and reflecting (b) superposition eye. Both redirect the rays as required by Fig 8.3.

'Rays from an object point entering through different facets are superimposed not by refracting systems as in other superposition eyes but by a radial arrangement of orthogonal reflecting planes which are formed by the sides of the crystalline cones and the purine layers surrounding them.' As Fig. 8.14 shows, the mirrors direct light to a common focus. Mirrors are inverters, just like the telescopes in refracting superposition eyes (Fig. 8.3b), and so the ray-bending that the two kinds of optical element perform is almost identical. However, problems start to arise when one tries to work out what will happen to rays that are not in the idealized central plane shown in Fig. 8.14b. In general, rays in oblique planes will not encounter just one side of each mirror box, but two. What happens to such rays? Do they, like the singly reflected rays in Fig. 8.14, all reach a common focus?

It turns out that the square arrangement of the facet array (almost unique to the decapod crustaceans) is crucial here. The principle is that of the 'corner reflector'. Corner reflectors – two mirrors at right angles – are occasionally encountered in hairdressers and clothes shops, where they have the disconcerting property that wherever you move they continue to reflect back your image. The reason for this peculiar property is shown in Figure 8.15a. A ray reflected from the two mirrors must be rotated through a total of two right angles, which means that it will return parallel to its original direction, *no matter what angle the ray initially makes with the mirror pair.* In other words, apart from a slight lateral displacement of the reflected ray, a corner mirror behaves as though it were a single mirror, but one that is always at right angles to the incoming ray. This property turns out to be very useful, for example in radar reflectors for ships and buoys, and it is also the property that makes reflecting superposition possible.

Consider first an arrangement for producing a point image by reflection that does not involve corner reflectors (Fig. 8.15c, left-hand side). This consists of a series of concentric 'saucer rims', each angled to direct rays to a common focus; Fig. 8.14b would then be any radial section through this array. The problem here is that such an array of bands has a single axis, and only rays nearly parallel to that axis form an image; other rays are reflected chaotically around the stack. The alternative is to replace the single reflecting bands with an array of corner reflectors – two sides of each mirror box (Fig. 8.15c, right-hand side). This substitution is possible because each corner behaves as though it were a single, appropriately oriented, mirror. Rays behave almost as though they had encountered a mirror strip in the saucer rim array. However, the beauty of the corner-reflector arrangement is that the orientation of each mirror pair is no longer important, unlike the situation in the single mirror array. Thus, the structure as a whole no longer has a single axis and can be used to make a wide-angle eye. Clearly, this mirror-box design only works with right-angle corners and not hexagons, which accounts for the square facets (Fig. 8.13).

Various other features of these eyes are important for their function. The mirror boxes must be the right depth, two to three times the width, so that most rays are reflected from two of the faces, but not more. Rays that pass straight through are intercepted by the unsilvered 'tail' of the mirror boxes, and Vogt (1980) showed that its refractive index decreases in such a way that appropriate critical angle reflexion continues to occur through the clear zone. Finally, there is a weak lens in the cornea of the crayfish. This lens 'pre-focuses' the light that enters the mirror box, thus giving a narrower beam at the retina. All these features provide an image generally comparable in quality to that produced by refracting superposition optics (Bryceson and McIntyre 1983), although it does seem that rays which make too many or too few reflections contribute to measurable stray light (glare) in the image on the retina.

Given the ancient origin of the decapods, the reflecting superposition mechanism presumably evolved within that group back in the Cambrian. Interestingly, the larval stages of decapod shrimps have apposition eyes with hexagonal facets, which change at metamorphosis into superposition eyes with square facets (Nilsson 1989). This transformation strongly suggests that the apposition eye is ancestral, and that the development of reflecting superposition occurred as a brighter image was needed for dimmer, benthic, life. The retention of apposition eyes into adult life by the brachyuran crabs, normally regarded as 'advanced' decapods, no doubt reflects the crabs' littoral or semi-terrestrial environment, in which light levels are generally high.

Parabolic superposition

This final type of eye is the most recently discovered (Nilsson 1988) and the most difficult to understand. From an evolutionary viewpoint, it is also the most interesting because it has some characteristics of apposition eyes, as well as both

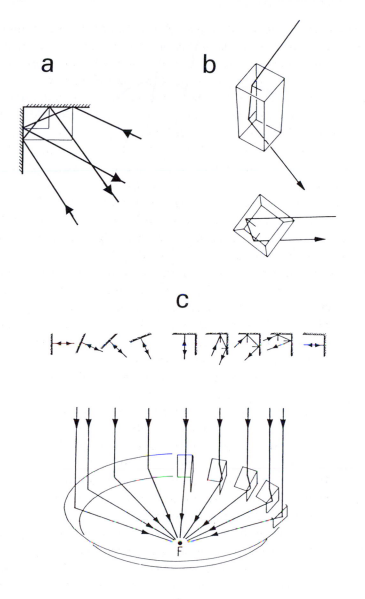

Fig. 8.15 (a) The principle of the corner reflector. Whatever the angle of incidence, incoming rays are reflected through two right angles, and so emerge parallel to their original path. (b) In the mirror box of a shrimp, the incident and reflected rays are not in the same horizontal plane, but viewed along the axis of the structure (*below*) the corner reflector behaviour is evident. (c) Ray paths for rays that are not in the central plane of the eye (unlike Fig. 8.14b). The mirror boxes act as 'corner reflectors' in which rays reflected from two sides emerge in the same plane as the incident rays to reach a focus at *F*. This is the condition for obtaining a clear image over a wide angular field. See text for further explanation. *C* is the eye's centre of curvature.

Fig. 8.16 Parabolic superposition. *Left*: rays are focused by the cornea to a point near the bottom of the crystalline cone. However, oblique rays are intercepted by the silvered walls of the cone and redirected back across the axis to form a beam contributing to the superposition image (see Fig. 8.3 and 8.4). *Right*: view from above. In this plane rays are focused more strongly onto the wall of the crystalline cone, and then brought parallel again by the same cylindrical lens. In both planes a parallel input beam emerges as a parallel output beam.

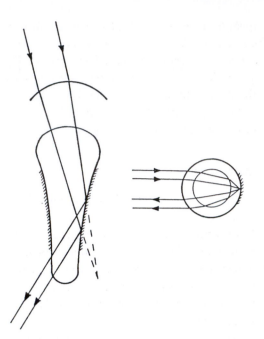

other types of superposition eye (Fig. 8.16). It was first discovered in a swimming crab *(Macropipus = Portunus)*. Each optical element consists of a corneal lens, which on its own focuses light close to the proximal tip of the crystalline cone, as in an apposition eye. Rays parallel to the axis of the cone enter a light-guiding structure that links the cone to the deep-lying rhabdom. Oblique rays, however, encounter the side of the cone, which has a reflecting coating and a parabolic profile. The effect of this mirror surface is to recollimate (make parallel) the partially focused rays, so that they emerge as a parallel beam that crosses the eye's clear-zone, as in other superposition eyes. This relatively straightforward mechanism is complicated because rays in the orthogonal plane (perpendicular to the page) encounter rather different optics. For these rays, the cone behaves as a cylindrical lens, thus creating a focus on the surface of the parabolic mirror. The same cylindrical lens then recollimates the rays on their reverse passage through the cone (Fig. 8.16, right). This mechanism has more in common with refracting superposition. Thus, this eye uses lenses and mirrors in both apposition and superposition configurations and it would be the ideal ancestor of most kinds of compound eye. Sadly, the evidence is against this, as all the eyes of this kind so far discovered in crustaceans are from the brachyuran crabs or the anomuran hermit crabs, neither of which is an ancestral group to other crustaceans (Nilsson 1989). However, this eye does demonstrate the possibility of mixing mirrors and lenses, thus providing a viable link between the refracting and reflecting superposition types. This is important because such transitions do appear to have occurred. The decapod shrimp *Gennadas,* for example, has a perfectly good refracting superposition eye, whereas its ancestors presumably had reflecting optics as in related shrimps (Nilsson 1990). A variant of parabolic superposition, which uses square mirror boxes with parabolically tapering sides rather than cylindrical lenses, occurs in Xanthid crabs and Atalophlebid mayflies. The latter provide the only known example of a mirror-based superposition mechanism in insects.

Summary

1 Superposition eyes produce real, erect images on a retina separated from the optical elements by a clear zone.

2 In refracting superposition eyes the optical elements may be lens cylinders or corneal lens/lens cylinder combinations. These act as inverting telescopes.

3 Resolution can be almost as good as the diffraction limit allows, and the sensitivity is usually much greater than in an apposition eye of the same size. Double eyes, with different resolution in the two parts, occur in both insects and crustaceans.

4 Superposition eyes often exhibit eye glow, when they are illuminated from the viewing direction. This results from a reflecting tapetum behind the retina.

5 Butterflies have afocal apposition eyes. This system is closely related to refracting superposition, except that the telescopic elements have a much higher magnification than those of moth superposition eyes. Light enters the rhabdom as a parallel beam, rather than as a focused image as in ordinary apposition eyes.

6 Shrimps, crayfish, and lobsters have superposition eyes in which the optical elements are not lenses but mirrors. The reflecting surfaces are at right angles to the eye surface, and form a square array. Most rays encounter two faces of each square, and this corner-reflector configuration makes it possible for the eye to form an image over a wide field of view.

7 A third mechanism, parabolic superposition, makes use of a lens/mirror combination to form the dog-leg ray path necessary for superposition imagery. This is found in certain crabs.

9 | Movements of the eyes

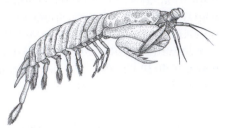

Movements of the eyes

Most of the chapters in this book have been concerned with the ways that eyes produce images, and how these images are sampled by the retina. This approach gives the impression that eyes are static devices which register the scenes in front of them rather like surveillance cameras, recording the positions and motions of objects within a fixed field of view. This, however, is a very misleading picture of the way that eyes deal with the world, because most animals with good eyesight have mobile eyes and images that change moment by moment. They move either because the animal they are attached to moves in the world, or because the head moves on the body, or because the eyes move in the head. In humans and many other animals all three kinds of motion contribute to eye movement, and the field of view of the eyes is rarely still for more than a few tenths of a second at a time. The eyes are not simply dragged around by the platform they are attached to. In primates particularly vision is a very active process; our eyes search the sur- roundings for information rather than simply absorbing it. To complete our account of the way that eyes work we need to explore this dynamic aspect of their operation. Eyes sample in time, as well as space.

We will first examine the nature of our own eye movements and ask how they contribute to vision. This is followed by a brief survey of the eye movements of other animals, to see whether there are common patterns across the animal kingdom. It turns out that there are, and this leads to the next question: why this should be so? A major component of the eye movement strategies of most animals is gaze stabilization. For example, a clockwise head movement is typi- cally accompanied by an anti-clockwise eye movement, so that although the eyes are seen to move relative to the head, what they are *really* doing is keeping the image on the retina stationary, despite motion of the head or body. Paradoxically, eye movements are just as concerned with keeping gaze still as they are with changing its direction. The princpal reason for this is that the photoreceptors themselves are quite slow: it takes 10 milliseconds or more for a receptor to respond fully to a change in light intensity, and this means that changes in the image that occur faster than this are lost. Just as in photography, where it is important to avoid blur by keeping the camera still, so with eyes. Motion degrades images, and the various systems that stabilize the eyes are organized to prevent this happening. Finally, we discuss some interesting exceptions to these rules, animals with scanning eyes that sweep their gaze across the scene in a manner that our eye-movement system simply prohibits. Why are they doing this, and how do they get away with it?

How humans acquire visual information

When we look at a scene we have the impression that it is stationary with respect to our viewing point, and that we see all parts of it with full clarity and resolution. If something or someone moves within the scene we see that quite appropriately as motion in the world 'out there', but we see very little evidence of motion brought about by our own movements: of eyes, or head, or body. We may be dimly aware that we 'pan' gently around a scene: 'She let her eyes wander over his …'. But this is an illusion. Our eyes take in the scene before us in a staccato barrage of saccades. These are the brief fast eye movements that convey the centre of gaze from point to point with a frequency of up to three per second. Our eyes do not wander at all: they jump around! Although these movements have been known about for well over a century, it was not really until a famous series of illustrations of eye movements across scenes were published by Alfred Yarbus (1967) in his book *Movements of the eyes* that this quite counter-intuitive notion of how we view the world became compelling. Between saccades we have periods of nearly stationary viewing – fixations – that last for about 300 ms, or longer if our attention is caught, and this 'fixate and saccade' strategy seems to be our main way of doing visual business with the world.

Figure 9.1 shows how this strategy works in practice. When doing real tasks (in this case filling a kettle prior to making a cup of tea) saccades are aimed at points in the surroundings from which visual information is needed to execute the job in hand. So the eyes go to the kettle, then the sink, the kettle lid, the taps, and then the water stream, as the changes in the task require as it progresses. Notice that this is not a 'random walk'. Behind the scenes the parts of the brain responsible for eye movements 'know' just what is required of them by the motor program the brain is trying to execute.

Intriguing though the question is, we will not discuss here the reasons why we do not see our own saccadic eye movements. That problem has remained essentially unresolved for more than a century. Instead we will look at why it is that we have this saccade and fixate strategy in the first place. An idea that comes to mind immediately is that the fovea, being small (in angular terms it is only about 1° across, the subtense of a thumbnail at arm's length) has to be redirected

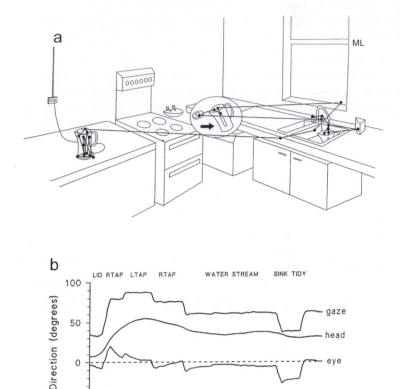

Fig. 9.1 The human saccade and fixate strategy in action. (a.) Record showing the first 25 fixations made by the author when filling a kettle prior to making a cup of tea. Dots are fixations, lines are the paths of saccades. From Land, M.F., Mennie, N. and Rusted J. (1999). (b.) The roles of eye and head movements in fixation sequences. The records show horizontal eye rotations relative to the head (eye), head rotation in space (head), and gaze rotation in space (gaze = head + eye) for the final series of fixations in Fig. 9.1a, from the kettle lid to the water stream. Notice that the eye record contains both fast saccades and slow movements that are the exact opposite of the head movements (the vestibulo-ocular reflex). The result is that gaze fixations are steady, and unaffected by head movements. Dashed line shows the straight ahead direction of the eyes; the other two traces have been arbitrarily displaced on the ordinate for clarity.

from place to place in order to give us the high resolution information we need from different parts of the scene. Whilst this is certainly crucial for primate vision, we have to remember that most vertebrates do not have foveas, and yet they too use the same saccade and fixate strategy. It was Gordon Walls, famous for his book *The vertebrate eye* who first pointed out that the reason why our early fishy ancestors adopted this strategy was to keep gaze still during locomotion, so that surroundings could be seen without motion blur.

Box 9.1 Reflexes that stabilize the eye

In vertebrates the eyes are prevented from involuntary rotation relative to the surroundings by two reflexes, the vestibulo-ocular reflex (VOR) and the opto-kinetic reflex (OKR). They both cause the eyes to counter-rotate in their sockets when the head rotates, thus keeping the retinal image more or less stationary.

The vestibulo-ocular reflex is driven by the rotation detectors in the semicircular canals of the inner ear (Fig. 9.2). When the head rotates the fluid in these canals lags behind the surrounding structures. Mechano-sensitive hair cells, attached to a jelly-like cupula protruding into the fluid, are bent by the relative motion of the fluid, and fire action potentials in proportion to head rotational velocity. This signal is received by the vestibular nuclei and passed on to the oculomotor muclei that innervate the eye muscles, with the result that the eye moves in the opposite direction to the head rotation. The six eye muscles, operating in antagonistic pairs, provide stabilization about all three axes. Although the signal provided by the vestibular system is one of velocity, the signal to the eye muscles is essentially one specifying the rotational position of the head. This means that somewhere in the system the signal is integrated from velocity to position, and the exact location of this integrator has been a subject of much interest. VOR is not a feedback system; the eye movements themselves do not affect the semicircular canals. This means that (like throwing a dart) the system needs to be well calibrated, and to have a gain (eye-movement size/head-movement) of exactly –1. Interestingly, divers have problems because their visual world, seen through an air-filled mask, moves faster across the retina than the speed at which the head rotates (by a factor of 1.33, the refractive index of water). They find difficulty in living with two gains for the VOR, one for land and another for under water.

The optokinetic reflex is a feedback loop which uses signals from motion detectors in the retina to activate the eye muscles which thus 'null out' any residual movement between image and retina (Fig. 9.2). If the image moves clockwise across the retina, a clockwise movement of the eye itself will decrease the relative motion. (Note that an *anti*-clockwise rotation of the head will cause the same image movement). The real function of this reflex is to clamp the eye to the visual world, which (barring earthquakes) can reasonably be supposed to be stationary. Typically, however, this response is studied by placing the subject (human or animal) inside a rotating striped drum (see Fig. 4.2). The eyes (and/or head and body) then follow the stripe pattern, and this generally leads to a pattern of eye movements known as *nystagmus*. The eyes follow the stripe pattern for a while, and then flick back to a more central position before resuming the following motion. The resulting sawtooth-like behaviour is said to have fast (resetting) and slow (following) phases. Because

Box 9.1 Reflexes that stabilize the eye (*contd.*)

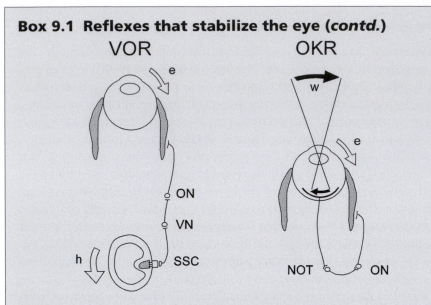

Fig. 9.2 Diagrams showing how the vestibulo-ocular reflex and the optokinetic reflex stabilize the eyes. In the vestibulo-ocular reflex the semicircular canals (SSC) measure rotational head velocity (*h*). This signal is relayed via the vestibular nuclei (VN) to the oculomotor nuclei (ON) which innervate the eye muscles. The result is a movement of the eyes (*e*) equal and opposite to the head movement. In the optokinetic reflex any movement of the image, whether brought about by movement in the world (*w*), or by movement of the head, is detected by ganglion cells in the retina. The signal passes to the nucleus of the optic tract (NOT) and thence back to the eye muscles via the oculomotor nuclei. The result is a cancellation of the original image motion. Note that, in this feedback loop, what the eye sees is an error signal, the 'slip' speed across the retina (*w–e*). This has to be amplified in the brain to provide an eye speed comparable with the original external disturbance.

of the way the behaviour is evoked, it appears that its function is to cause the eye to pursue moving targets, and this has led, historically, to considerable confusion. Under normal circumstances the velocity of the surroundings is zero, and as the system operates to minimize the velocity of the image on the retina, the result will normally be a stationary eye. To confuse matters further, humans and other primates *do* have true pursuit behaviour, but this operates only for small foveated targets; the optokinetic response involves the whole image, and actually has to be over-ridden when a small target is to be tracked.

The two stabilizing reflexes operate over different velocity ranges. OKR is slow, and for oscillating backgrounds is only effective up to about 1 Hz. VOR, on the other hand, operates up to 10 Hz (it is quite difficult to dislodge gaze by shaking your head) but fails at very low frequencies. Between them the two reflexes keep the image almost stationary over the whole range of rotational speeds likely to be encountered by a moving animal.

'Their origin (eye movements) lies in the need to keep an image
fixed on the retina, not in the need to scan the surroundings.'

(Walls, 1962)

Very early in the vertebrate lineage the powerful vestibulo-ocular and optokinetic reflexes evolved, whose function was to stabilize the eye against movements of the head (see Figs 9.1b, 9.2, and Box 9.1). However, as an animal turns
as it moves through the environment, stabilization alone is not enough; the eyes
must move to re-centre gaze from time to time or they will finish up in one or
other extreme position. Saccades are the means of achieving this. Their impressive speed reflects the need to keep the time they blur the image to a minimum.
Humans spend about 10 per cent of their waking hours engaged in saccades,
during which vision is degraded, either through blur or 'saccadic suppression'
when vision is actively suppressed. Amazingly, this amounts to about one and a
half hours of near blindness each day.

Humans and other primates, but probably not many other vertebrates, have
the ability to track objects smoothly, provided they do not move too fast or too
unpredictably. Smooth tracking, or pursuit, is more than just a fixation on a
moving target. Numerous studies have shown that the pursuit mechanism has a
sophisticated control system capable of anticipating the motion of objects, when
this is at all predictable. The system also has the capacity to separate moving
foreground objects from the stationary background. Indeed, it is necessary to
suppress the effect of background motion, as this normally contributes to the
optokinetic response whose function is to prevent relative motion of retina and
image. In smooth tracking that clamp has to be removed.

Are other animals like us?

We may ask whether the 'saccade and fixate' pattern of eye-movement behaviour, just outlined, is merely a phylogenetic quirk of the vertebrate lineage, or
whether the same reasons for keeping the image still apply to vision in all
animals with good eyesight. Table 9.1 gives a brief classification of human eye
movements that can serve as a basis for making comparisons across phyla (an
excellent text is provided by Carpenter 1988).

Figure 9.3 shows recordings of the movements of the eyes of a fish and a crab,
both of which are engaged in locomotion that involves some rotation. Taking the
goldfish first, the records show the head rotating relatively smoothly through
about 100°. The record of the eye movements relative to the head (R EYE/HEAD),
however, is quite different. There are a number of fast saccadic movements in the
same direction as the head movement, and between these the eyes rotate
smoothly in the opposite direction to the head. The sum of the head-in-space and
eye-in-head movements gives the movements of the eye-in-space, i.e. movements
of gaze: these are shown in the top trace. The result is a series of periods of sta-

Table 9.1 Types and rôles of human eye movements

FAST (saccades)	1. 'Voluntary' relocation of the direction of gaze
	2. Targeting new stimuli in the periphery
	3. Fast phases of optokinetic nystagmus (recentring movements)
SLOW	1. Compensatory movements that stabilize gaze (vestibulo-ocular reflex and optokinetic reflex)
	2. Foveal tracking of small targets
	3. Vergence movements (tracking in depth)

(Microsaccades, drift, and tremor also occur. These small movements of a few minutes of arc may serve to prevent the image from fading. However, it seems that there is always enough residual image motion from other sources to keep the image 'refreshed'.)

tionary gaze, with fast saccades that shift the gaze direction from time to time through angles of 10–30°. This is the 'saccade and fixate' strategy mentioned above in connection with our own vision. Perhaps it is no surprise to find that we are rather like fish; after all we share a common ancestry. The rock crab's vision, however, evolved quite independently from that of vertebrates, and crabs have a quite different design of eye (apposition, see Chapter 7). Nevertheless, the pattern of head, eye, and gaze movements is remarkably similar to that of the goldfish. Again, the head (part of the body in a crab) rotates relatively smoothly, whilst the eyes execute both fast saccades and counter-rotations (the angular scale is magnified here) and the resulting gaze movements are fast refixations alternating with stationary periods.

In insects the eyes are physically part of the head and do not move relative to it. A head movement for a fly is thus an eye movement as well. This is shown in

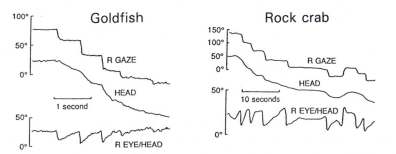

Fig. 9.3 Eye, head, and gaze movements during curvilinear locomotion in a goldfish and a crab. In both cases gaze is actively stabilized against movements of the body, except during fast (saccadic) gaze changes. Goldfish record from Easter, Johns, and Heckenlively (1974). Crab record from Paul, Nalbach, and Varjú (1990).

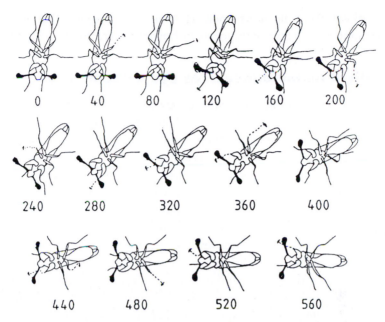

Fig. 9.4 A stalk-eyed fly turning through 90° on a pane of glass. Notice that the body rotates continuously and smoothly, but the head makes the turn in two saccadic jumps, one at 120 ms and the second at 380 ms. Between the saccades the head counter-rotates relative to the body.

Fig. 9.4 in which a stalk-eyed fly (chosen because the stalked eyes make the head movements more easily visible) rotates through 90° on a glass plate. Notice that the body movement is continuous, but that the head and eyes rotate in a series of saccade-like turns, one after 120 ms, and another after 380 ms. Between these saccades the head is actually counter-rotating relative to the body. Even without independently mobile eyes, the fly is doing exactly the same thing as the fish and the crab: keeping gaze still as the body rotates. Thanks to a remarkable study by Schilstra and van Hateren (1998) we now know that flying flies show the same kinds of movements.

Rather like the flies, birds also make head saccades, and these are a very obvious feature of their visual behaviour. Birds do have eye movements, but their main function seems to be to 'sharpen up' the head saccades. As the head turns the eyes counter-rotate briefly, then flick to the new position and again counter-rotate until the head comes completes its saccade. But the eyes do not seem to make saccades on their own. Head saccades can also occur in humans. A particularly interesting case arose recently of a woman who has no eye movements, due to a rare fibrosis of the eye muscles, and yet is able to read fluently and conduct her life more or less normally (Gilchrist *et al.* 1997). In reading she makes rapid head movements that, although slower than typical eye saccades, serve the same function, and when she is engaged in everyday tasks these give an extraordinarily bird-like impression. This reinforces the view that the saccade and fixate strategy is of crucial importance, however it is achieved.

Insect flight behaviour seen as eye movement

For a light insect not attached to the ground there is nothing to prevent it using body manoeuvres to move its eyes. This is what happens with hoverflies (Syrphidae), whose superb mobility makes even neck movements superfluous. A good example is the small hoverfly *Syritta pipiens*. Female flies hover around flowers, feeding on nectar, whilst the males spend much of their time in stealthy pursuit of the females. The males have an advantage in that they have an 'acute zone' in the front-facing part of the compound eye, where the resolution is about three times better than anywhere in the female eye (Chapter 7, Fig. 7.18a). Thus the males can shadow the females around until they land, whilst remaining effectively out of sight. Figure 9.5 shows an example of this. It is clear that the flight behaviour of the female (above) and male (below) are not the same. Although the female's flight is continuous, her turning is not. She makes rotational saccades from time to time (e.g. just before 3, just after 5) and between these the body does not rotate, even though translational flight (non-rotational

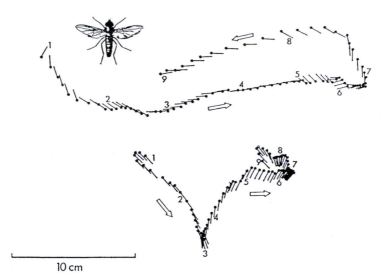

10 cm

Fig. 9.5 Tracking behaviour by the hoverfly *Syritta pipiens* filmed from above. The record shows the flight path of a female (*above*) being tracked by a male (*below*). Notice that the female's flight has a 'saccade and fixate' pattern (see Figs 9.3 and 9.4), with very little rotation between saccades, although there is no restriction on translation. The male, however, tracks the female smoothly, keeping her within 5° of his midline. Here his eye has a region of high acuity, absent in the female. Also notice that the male maintains a constant distance from the female. Corresponding instants are numbered every 400 ms. From Collett and Land (1975).

movements of the whole body) may occur in any direction. The flight of non-tracking males is similar. As soon as they begin to track, however, the pattern changes dramatically. Throughout the 3.6s period shown in Fig. 9.5 the male points directly towards the female, tracking smoothly, and keeping her within the ± 5° forward sector containing his acute zone. Notice, too, that he maintains a roughly constant distance of about 10 cm, which is important if he is to remain undetected. Interestingly, if the female moves fast he switches to a saccadic mode of tracking, just as we do. Amongst insects, *Syritta* shares an ability to track both smoothly and saccadically with praying mantids (Rossel 1980), although in the latter the tracking is mainly achieved with movements of the head, rather than the body.

The examples in the last two sections make it clear that the human oculomotor system is far from unique in the way it samples the world around us. Many, perhaps most, animals with good eyesight do something similar, although this may not always present itself as movements of the eyes in the head. We now inquire what the reasons for this consensus might be.

Why not let the eyes wander? Some consequences of image motion

What are the arguments against allowing continuous image motion, and in favour of sampling via a series of more or less stationary images? Three seem the most persuasive:

A. Resolution is lost if motion blurs the image.

B. It is easier to see motion of external objects if the retinal image of the background is stationary.

C. It is easier to obtain heading and distance information from the pattern of motion on the retina that results from locomotion, if the rotational motion has already been removed.

We will consider each of these in turn.

A. Motion blur

There are good reasons why fast motion must degrade an image, and they have to do with the rather slow rate at which photoreceptors respond to changes in intensity. Figure 9.6a shows the response of a locust photoreceptor to a brief flash of light. When light-adapted the response takes about 20 ms (milliseconds) to reach its peak, and it has a duration of about 30 ms, if we ignore the end of the tail. By 'cumulating' the curve (adding it to itself but shifted in time) we can work out how large the response would be to sustained light pulses of different

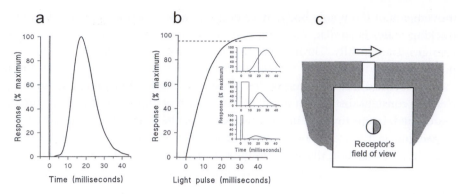

Fig. 9.6 The blurring of the image that results from image motion. (a.) The electrical response of a light-adapted locust photoreceptor to a brief flash of light. It takes about 40 ms to complete (from Howard *et al.* 1984). (b.) Main graph shows how the response to a pulse of light increases in size as the pulse lengthens, reaching 95 per cent of its maximum value when the pulse is 25 ms long. Insets show the responses to pulses of 2, 9, and 20 ms. (c.) A light stripe moving across the field of view of a receptor provides the receptor with a pulse of light whose duration depends on the stripe's speed. If this pulse is shorter than 25 ms [see (b)], its intensity will not be accurately signalled, and the result will be seen as a blurred image.

durations. As Fig. 9.6b shows, the response is very small for short pulses of light (lowest insert), but reaches 95 per cent of its maximum, i.e. the plateau value for a sustained stimulus, when the pulse is 25 ms long. No further increase is seen for pulse durations much longer than 30 ms. Thus this receptor needs about 25–30 ms of light if it is to signal the change in intensity fully.

Now suppose that the change is not brought about by turning on a stationary light, but by a moving bright band in the environment, imaged onto the receptor (Fig. 9.6c). The band we have chosen has the same angular width as the receptor's acceptance angle, because this is the narrowest band that will provide a full signal to the receptor when it is stationary; there is no degredation due to the receptor's diameter alone. If the stripe moves so fast that it illuminates the receptor for only, say, 5 milliseconds, then from Fig. 9.6b the response it will evoke is only about 30 per cent of the maximum value. On the other hand if it illuminates the receptor for 30 milliseconds it will produce a full response. If the receptor's field of view is 1° across, this means that the maximum speed the stripe can move, and still generate a full response, is 1° in 30 ms, i.e. 33°s^{-1}. Notice that if the acceptance angle and the stripe had both been 5° wide instead of 1°, but the response time stayed the same, then a full response would be generated up to speeds of 5° in 30 ms, i.e. 167°s^{-1}. In other words, poorly resolving systems with large receptor acceptance angles (high $\Delta\rho$) can tolerate higher velocities than better resolving eyes. The loss of response to images that move faster than the permitted limit is what we know as motion blur. It shows itself first as a loss of contrast in the highest spatial frequencies in the image (see Land 1999). We can generalize the results in this paragraph into a useful rule of thumb, as follows. *Significant blurring of the image occurs at angular velocities that exceed one receptor acceptance angle per response time*. We will refer to this as the 'blur rule' in the rest of this chapter.

In humans the acceptance angle of foveal cones is about 1 arc minute, and the response time about 20 ms, which implies that blurring will occur at image speeds of close to $1°s^{-1}$ for high spatial frequencies. This fits well with psychophysical studies showing that when a pattern moves at only $3°s^{-1}$ across the retina all high spatial frequency information (greater than about 8 cycles/deg) is lost. $3°s^{-1}$ is not very fast, and this explains why we need to stabilize vision so tightly, using the vestibulo-ocular and optokinetic reflexes. Interestingly, vision also fades when the image is kept very well stabilized, and so some low-speed image motion is actually desirable (see the comment on Table 9.1). Very few species have spatial resolution as good as ours. In particular, insects have much wider receptor acceptance angles than we do (~1° rather than 1 minute), so they can be much more tolerant of image motion. Their receptors tend to be somewhat faster too (response time down to less than 10 ms for a fly in daylight), making the blur rule limit about two orders of magnitude greater than in humans. Thus insects such as bees and flies should be able to tolerate angular velocities of at least $100°s^{-1}$ without significant resolution loss. Again, this conclusion is well supported by both behavioural and electrophysiological evidence.

B. Movement detection

Many arthropod species do not allow their gaze to drift to anything like the extent that the blur rule suggests. Hoverflies, for example, do not allow their bodies to move even a few degrees per second, when they are hovering in wait for passing females. Bombyliid flies and some solitary bees have a similar capacity for remaining rotationally still. Recently, Layne *et al.* (1977) found that walking crabs stabilize their eyes about 10 times better than they need to in order to preserve vision from blur. On the other hand, honey bees and common wasps seem to meander about in a much less tightly constrained way (see Fig. 9.13).

Hoverflies in particular need to detect small moving objects, usually against a background of foliage, and it is easy to imagine that this is a particularly demanding task. To our knowledge there is no theoretical account of why a stationary eye can detect motion better than a moving one, but there are some human psychophysical studies that bear on this. Relative motion between two surfaces is easy to detect when one of the two is nearly stationary, but detection becomes rapidly more difficult when both move, even though the difference in velocity remains the same. The threshold for this increase is about $0.3°s^{-1}$, one-tenth the threshold for acuity loss due to blur.

If this is true, then keeping the eyes totally still should be a good way to ensure maximum detection of objects that move. Interestingly, this may be even more effective than one might expect because, in humans at least, a totally still image fades in a few seconds, and when this has occurred the only detectable objects are those that move. We cannot, in normal life, keep our eyes still enough to lose the image of the overall scene, but it may well be that other animals can. It is an attractive idea that when rabbits or squirrels hold their heads and eyes still they are able to see movement but nothing else, and similarly for toads, snakes, or wolf spiders. It simplifies the world enormously if the 'high-pass filter' nature of the early visual process can be used to restrict what can be detected to just those things that are of vital importance: those that move. There is, however, an unresolved paradox here in that the very stability of gaze, especially in an animal like a hoverfly trying to keep still in three dimensions, is itself dependent on optokinetic reflexes, and so in some sense the animal is still 'seeing'. The reflexive aspects of motion vision may operate by different pathways and with different rules from those concerned with the detection of behaviourally significant objects, but we have little real information to go on.

C. Disambiguating the flow-field

The third reason for preventing motion of the image brought about by rotation of head or body concerns the use of information in the retinal motion pattern (flow-field) generated by body movement. This is of two kinds (see Gibson 1979). When our eyes rotate, the whole field moves across the retina in a uniform way (Fig. 9.7b), but when we translate (i.e. move in a straight line) the pattern of motion is much more interesting (Fig. 9.7a). The point towards which we are moving is stationary on the retina, and motion radiates from it, making it the

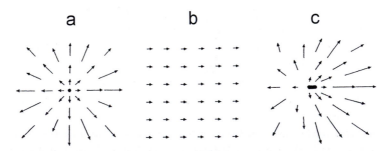

a   b   c

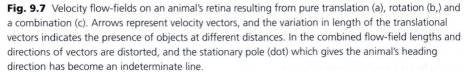

Fig. 9.7 Velocity flow-fields on an animal's retina resulting from pure translation (a), rotation (b,) and a combination (c). Arrows represent velocity vectors, and the variation in length of the translational vectors indicates the presence of objects at different distances. In the combined flow-field lengths and directions of vectors are distorted, and the stationary pole (dot) which gives the animal's heading direction has become an indeterminate line.

'focus of expansion' on the retina. The motion pattern reaches its maximum velocity to the side, and then, if we could see it, it contracts again behind us. This basic pattern is modulated by the distances of objects. When an animal's eye moves through space, near features move faster across the image than distant ones, and locomotion can thus generate distance information. This was discussed in Chapter 7 in connection with design of apposition compound eyes, and the principle is explained in Fig. 7.13. If the animal has an estimate of its velocity, then the pattern of angular velocities on its retina converts quite simply into a map of inverse object distances.

However, a serious problem is that animals rarely move in straight lines. There is always a certain amount of rotation, and as Fig. 9.7c shows this will corrupt the translational flow-field, making it difficult to interpret. The focus of expansion is no longer a point but a blurred line that cannot be used by the animal to determine its heading, and all the retinal vectors are distorted, making distance judgments harder. An effective cure for this is simply not to let the eye rotate, by applying gaze stabilizing reflexes (vestibulo-ocular and optokinetic). Then, between saccades at least, the eye will see an almost undistorted translational flow field. With rotation out of the way, retinal velocities can be read as distances.

An obvious question is whether or not there is any evidence that animals do actually use flow-field information to measure distance. One imagines that they must. Although animals have a great many ways of measuring the distances of objects (see Collett and Harkness 1982), few are available to a moving insect, which is essentially monocular and has a fixed focus eye. Simply not bumping into things requires a fast and accurate means of determining the three-dimensional layout of objects in the field ahead, and the translational flow-field is ideal for providing such information. In fact, there are surprisingly few studies

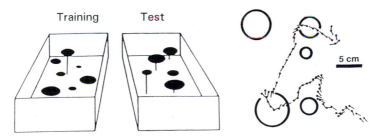

Fig. 9.8 The 'black meadow' experiment showing that bees can learn distances from the velocity flow-field (Srinivasan *et al.* 1989). Bees were trained to feed immediately above a black 'flower' of a particular height, and tested on a selection of flowers of different heights. The meadow had a transparent top that prevented the bees from flying at flower level. The bees chose the correct flower under conditions where the only cue to distance available to them was the motion parallax resulting from their own flight. The flight-path of a bee over the flowers is shown on the right; the bee tends to cross flower boundaries with straight-line segments of flight.

showing that flow-field information is used for distance judgment, but there is one very convincing demonstration in honey bees.

Lehrer *et al.* (1988) trained bees to fly over an artificial 'meadow' with black flowers of various sizes and of different heights (Fig. 9.8a). The bees were constrained to fly above the flowers at a constant height from the ground, which meant that the only available clue to the relative heights of the flowers came from retinal motion, and not from apparent size. The bees learned to associate a food reward with either high, low, or intermediate height flowers, showing that they had learned distances from the image velocity pattern generated during flight. Further evidence that this involved motion vision came from the finding that the system the bees were using was colour-blind and sensitive only to green contrast. The motion detecting system in bees is known to be sensitive only to green contrasts, unlike the trichromatic system usually involved in pattern recognition. Thus bees make explicit use of the distance information contained in the locomotor flow-field.

The bees appeared to be doing more than just picking up the distance information passively; they seemed to be trying to maximize the relevant input by orienting their flight appropriately, using straight flight segments to fly over the flowers, thus giving them a relatively uncontaminated translational velocity signal (Fig. 9.8b).

In the bee's case the ability to measure distance in this way comes as a useful by-product of normal locomotion, but other animals make much more deliberate parallax-generating movements. Before they jump, locusts frequently make side-to-side movements of their heads (peering). These movements are pure lateral translations, without rotation, which makes them ideal for registering the distances of objects by their image motion. Sobel (1990) found that moving the target as the locust peered generated artificial parallax, and that the locust's jump length was related to the actual image velocity, and no longer the actual distance. It is likely that this method of estimating distance, may be quite a common strategy. Young mantids make very similar peering movements, often

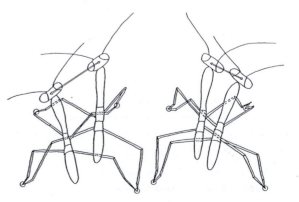

Fig. 9.9 Side-to-side scanning movements of an early instar praying mantis (*Sphodromantis lineola*). Note that the body moves in such a way that the head travels along a line that is almost exactly perpendicular to the animal's forward direction of view, and that the head does not rotate relative to the surroundings during a scan.

with a remarkable display of calesthenics (Fig. 9.9), which enable them to choose the nearest vertical object to jump to. Amongst vertebrates gerbils use a vertical head-bob for the same purpose (Goodale *et al.* 1990). A prerequisite for this behaviour is that the animal has the ability to keep track of particular edges in a cluttered environment, which implies quite sophisticated visual processing.

Of the three mechanisms discussed in this section, it seems that avoidance of blur (A) is the most basic because it applies to all animals with good vision, and the greater the acuity the better the stabilization must be. Detecting relative motion (B) is important for animals that need to see small moving objects, but is perhaps not a general requirement, and the use of the translational flow field for distance measurement (C) can perhaps be best thought of as a useful consequence of having the machinery for dealing with the first two. Whatever the mixture of reasons, however, one thing is clear: no animal should make smooth rotational eye movements, except in the special (and not very common) circumstance of tracking a moving target. Nevertheless, there are a few animals that break this taboo. As we shall see, however, it seems that they nevertheless manage not to violate the 'blur rule' given earlier. There are really two categories: those that have narrow retinae with which they scan the scene, and those with conventional retinae that push the limits of motion vision as far as the blur rule will allow. These will be considered in turn.

Exceptions: rotational scanning by one-dimensional retinae

In this section we will examine eyes that actively rotate in order to acquire information. Eyes of this kind are uncommon, but they do occur in several animal groups. The examples given here represent a series of increasing complexity. In all cases the relevant retinae are long and narrow, and operate more in the manner of industrial line-scan cameras than conventional cameras with two-dimensional images.

Prey detection by the sea snail *Oxygyrus*

The carnivorous planktonic sea-snail *Oxygyrus* is perhaps the most straightforward example of a scanning eye (Fig. 9.10). It has been known for a century that this group of molluscs have peculiarly narrow retinae, but how an eye with such a reduced field of view could be of use to the animal has only recently become apparent. *Oxygyrus* has a lens eye not unlike a fish eye, except that the retina is only 3 receptors wide by about 410 receptors long, and covers a field of about 3° by 180°. The one-dimensional structure of the retina would make very little sense unless it moved in some way, and indeed the eyes do scan (Fig. 9.10c).

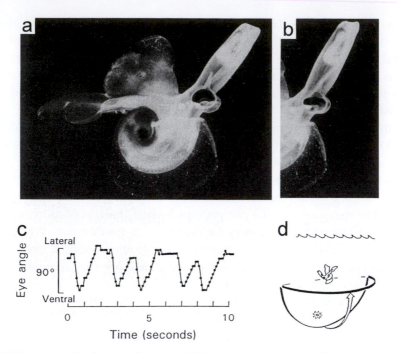

Fig. 9.10 Scanning with a linear receptor array. (a) Photograph of the sea-snail *Oxygyrus* shown with its eye pointing downwards (the snail does swim 'inverted' like this). (b) The appearance of the eye when directed laterally, a fraction of a second after (a). (c) The time course of 8 scans. The eyes move downwards very fast, and more slowly upwards. (d) Diagram showing the visual field of the eye during a scanning movement, and its probable role in detecting plankton. The retina is about 400 receptors long and 3 wide, giving a linear field of view which scans slowly upwards. Mainly from Land (1982).

The eyes move so that the retina sweeps through a 90° arc at right angles to its long dimension. The scanning pattern is a sawtooth, and the slower upward component has a velocity of $80°.s^{-1}$. The eye scans through the dark field below the animal, and the presumption is that it is searching for food particles glinting against the dark of the abyss.

Colour scanning in the mantis shrimp *Odontodactylus*

The mantis shrimps (Crustacea: Stomatopoda) are quite large crustaceans, only very distantly related to the more familiar decapod shrimps. Like their insect namesakes they are ambush predators, with a legendary ability to destroy their prey with smashing or spearing appendages (Plate 4). Their eyes are basically compound eyes of the ordinary apposition type, which provide an erect two-dimensional image. However, stretching more or less horizontally across each eye is a band of enlarged facets, 6 rows wide (Fig. 9.11 a and b). This mid-band,

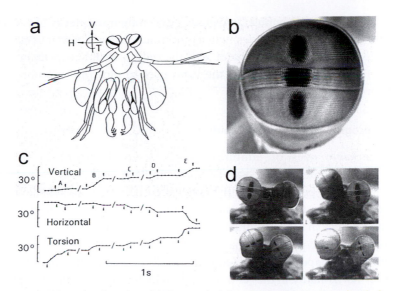

Fig. 9.11 An eye that scans for colour. (a) The mantis shrimp *Odontodactylus* has a band of ommatidia containing its colour vision system (black line) running across the centre of each apposition compound eye (see also Plate 4) . The vertical field of view of the band is very narrow – a few degrees. (b) Close up photograph of an eye showing the 6-row midband, and the three pseudopupils which indicate that there are three separate regions of the eye directed towards the camera. (c) Record of a series of small scanning movements. Notice that the eye rotates about all three axes [V, H, T on (a)], but by differing amounts. (d) Four photographs showing the eyes in a variety of positions. The eye movements are nearly always independent. The top two photographs show one or other eye with the 'acute zone' directed towards the camera, as shown by the wide pseudopupils (see Chapter 7). Mainly from Land *et al.* (1990).

which has a field of view only a few degrees in width, contains the animals' extraordinary colour vision system. This consists of four of the mid-band rows (the other two subserve polarization vision) and in each row the receptors are in three tiers. Each of these 12 tiers contains a different visual pigment, giving the animal the potential for *dodeca-chromatic* colour vision, with 8 pigments covering the visible spectrum, and a further four in the ultraviolet (Marshall *et al.* 1999). In adopting this impressive system, however, the mantis shrimps have set their eye movement system a daunting task. The outer parts of the eye operate as normal compound eyes – and are subject to the kinds of image stability considerations discussed earlier. The mid-band, however, has to move or it will not be able to register the colour of objects in the environment outside a very narrow strip. The result of this visual schizophrenia is a repertoire of eye movements unlike anything else in the animal kingdom (Land *et al.* 1990). In addition to 'normal' eye movements (fast saccades, tracking and optokinetic stabilizing movements) there is a special class of frequent, small (c. 10°) and relatively slow ($40°.s^{-1}$) movements (Fig. 9.11c), which give the animal a strange inquisitive appearance, perhaps because they resemble human saccades in their frequency of occurrence. They are, however, not saccades, which are much faster. These movements are typically at right angles to the band, and the only plausible explanation is that they are the scanning movements the band uses to pick out relevant coloured features in the surroundings.

Pattern recognition by jumping spiders

The remarkable eyes of jumping spiders were described in Chapter 5 (Figs 5.16–5.18, Plate 4). The secondary eyes (antero- and postero-lateral pairs) and large forward-pointing principal eyes have different roles in behaviour. The secondary eyes are fixed to the carapace and act only as motion detectors. If something moves in the surroundings these eyes initiate a turn, which results in the target being acquired by the principal eyes (Fig. 9.12). These eyes have narrow retinae shaped like boomerangs (Fig. 5.18), subtending about 20° vertically by 1° horizontally in the central region, which is only about 6 receptor rows wide. The high resolution was discussed in Chapter 5. Of interest here is the fact that the retinae of the principal eyes are moveable (the lenses themselves do not move). They can move horizontally and vertically by as much as 50°, and they

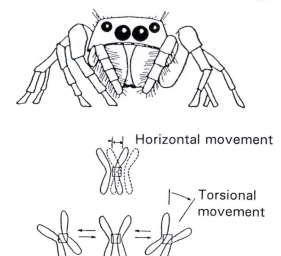

Fig. 9.12 The jumping spider *Phidippus*, showing the large movable principal eyes, and smaller fixed antero-lateral eyes (see also Plate 4). Below is a diagram and record of the movements of the boomerang-shaped retinae of the two principal eyes while scanning a novel target. These movements are conjugate, and consist of a stereotyped pattern of horizontal oscillations and slower torsional rotations. This scanning pattern apparently allows the narrow retinae to determine the angular pattern of edges in the target, and thus enables the spider to distinguish other jumping spiders from potential prey. From Land (1969).

Horizontal movement

Torsional movement

10° horizontal

Stimulus

10s

50° torsion

can also rotate about the optic axis (torsion) by a similar amount (see Land 1985). When presented with a novel target, the eyes scan it in a stereotyped way, moving slowly from side-to-side at speeds between 3 and $10°.s^{-1}$, and rotating through $\pm 25°$ as they do so (Fig. 9.12). We actually know what they are looking for: legs! In the 1950s Oscar Drees showed that jumping spiders are relatively indifferent to the appearance of potential prey, so long as it moves, but males are quite particular in what they regard as potential mates. Drawings consisting of a central dot with leg-like markings on the sides, however, will elicit courtship displays. Whatever its other functions may be, scanning in these spiders really seems to be concerned with feature extraction, the procedure itself apparently designed to detect the presence and orientation of linear structures in the target.

Scanning eyes: conclusions

In the first of these examples we have seen a one-dimensional retina used as a simple detector, rather like the scan line on a radar set. The other two are more interesting, however, because they combine more or less conventional two-dimensional eyes with special purpose line scanners, determining local colour in the mantis shrimp, and aspects of pattern geometry in the jumping spider. One cannot help feeling that the mantis shrimp solution, with both types of eye incorporated in the same structure, is an inherently clumsy one, because it forces the brain to time-share between the two systems. The eyes of the mysid shrimp *Dioptromysis* (Chapter 8, Fig. 8.10c), where highly 'foveate' vision alternates with wide-field vision, pose similar problems. In contrast, the dual system of fixed and moveable eyes of jumping spiders, with one set acting as target finder and the other as analyser, does seem to have much to commend it.

An obvious question raised by all these eyes is the extent of blurring that the scanning movements cause. Do they violate the blur rule limit, or not? It looks as though they are all just slow enough to stay within the rule. We do not know the response time of the receptors, but we know from other animals that 20 ms is a typical value for light-adapted eyes (Fig. 9.6a). If we make the assumption that the eyes are working at the blur limit, we can calculate a value for the response time (Δt) from the scan rate (s) and the receptor subtense ($\Delta \rho$), both of which are

Table 9.2 Scanning eyes: speed, receptor acceptance angle and estimated response time

Animal	Scan speed	Acceptance angle	Response time
	$(s)\ °s^{-1}$	$(\Delta \rho)\ °$	(Δt) ms
Labidocera (copepod)	219	3.5	16
Oxygyrus (mollusc)	80	1.1	15
Odontodactylus (stomatopod)	40	1.0	25
Metaphidippus (spider)	6.2	0.15	24

known ($\Delta t = \Delta\rho/s$). This has been done in Table 9.2 for the three species mentioned above, plus the copepod *Labidocera* which also has a scanning eye (Fig. 4.5c). The table shows that the calculated values for Δt are all in the range 15–25 milliseconds, i.e. close to the expected value, which means that the scanning rates are all close to the blur limit, but do not exceed it. Thus the animals are scanning as fast as they can without losing spatial information, which is presumably the optimal way to scan. The other interesting point is the inverse relationship, predicted by the blur rule, between resolution ($\Delta\rho$) and scanning speed. *Labidocera*, with poor eyesight, scans fast, but the jumping spider *Metaphidippus*, with excellent eyesight, scans very slowly.

Scanning procedures of hymenopteran insects

Anyone who has tried to eat a meal outdoors in August knows that wasps do not behave like hoverflies. They approach objects in an arcing path, apparently rotating continuously as they do so. The best studied examples of such behaviour are the orientation flights that wasps make both when they leave their nests before hunting for food, and immediately after they have located a new food source (Collett and Lehrer 1993). There is ample evidence that the function of these flights, which often have a remarkably consistent geometry, is to provide a framework for learning the landmarks of the local area. This provides both a means of recognizing the site on a return visit, and also a procedure for homing in on the goal. It is not our concern here to work out how this is done, but to see whether this behaviour is consistent with an unblurred image. Zeil (1993) filmed the nest-leaving orientation flights of wasps of the genus *Cerceris*, and measured both their rotational angular velocities, and also worked out the retinal angular velocities of the ground beneath the wasps during these flights (Fig. 9.13). He found that both sets of angular velocities showed a single fairly narrow peak between 100 and $200°.s^{-1}$. Similar velocities can be derived from the Collett and Lehrer's study of *Vespula*. The acceptance angles of the rhabdoms of these wasps have not been measured, but other similar insects have acceptance angles in the range of 1–3°, and a response time of about 10 ms seems likely for a hymenopteran in daylight. These assumptions give a blur limit in the right range ($100–300°.s^{-1}$), so we can reasonably assume that the wasps' strategy is to rotate as fast as is consistent with retaining the highest spatial frequencies in the image on the retina. These velocities would be quite intolerable for a human, but they are perfectly all right for an insect whose acuity is a hundred times worse. Insects, however, are not likely to make their already poor eyesight worse still, by permitting further motion blur.

As an afterthought, it may be that the reason why wasps are so apparently indifferent to the flailing of arms, napkins and kitchen utensils that their presence often induces, is that their rotation is fast enough to make the detection of rela-

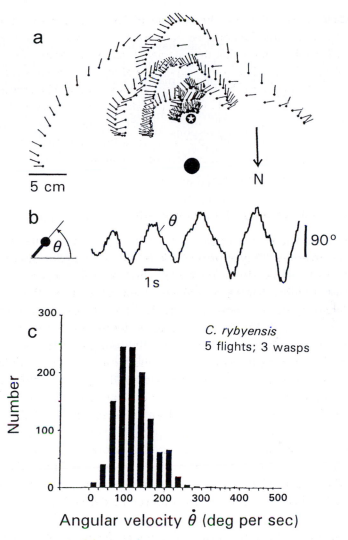

Fig. 9.13 (a) The flight path of the wasp *Cerceris rybyensis*, on leaving its nest, seen from above. The wasp is shown every 40 ms as it retreats from the nest in a series of arcs. The star is the nest and the large dot is a landmark. (b) Change of orientation (θ) during the flight in (a) showing the rather constant angular velocities in each direction. (c) Histogram of angular velocities during several such flights. All modified from Zeil (1993).

tive motion difficult, in keeping with the human psychophysical result discussed earlier. This contrasts with the way that animals such as male syrphids keep gaze very still indeed, presumably because the detection of the movements of external objects is exactly what they are trying to achieve.

Summary

1 Nearly all animals with good vision have a repertoire of eye movements. The majority show a pattern of stable fixations with fast saccades that shift the direction of gaze. These movements may be made by the eyes themselves, or the head, or in some insects the whole body.

2 The main reason for keeping gaze still during fixations is the need to avoid the blur that results from the long response time of the photoreceptors. Blur begins to degrade the image at a retinal velocity of about 1 receptor acceptance angle per response time.

3 Some insects (e.g. hoverflies) stabilize their gaze much more rigidly than this rule implies, and it is suggested that the need to see the motion of small objects against a background imposes even more stringent conditions on image motion.

4 A third reason for not allowing rotational image motion is to prevent contamination of the translational flow-field, by which a moving animal can judge its heading and the distances of objects.

5 Some animals do let their eyes rotate smoothly, and these include some heteropod molluscs, mantis shrimps, and jumping spiders, all of which have narrow linear retinae which scan across the surroundings. Hymenopteran insects also rotate during orientation flights at speeds of 100–$200°s^{-1}$. This is just consistent with a blur-free image, as are the scanning speeds of the animals with linear retinae.

Principal symbols used in the text

A	aperture diameter of a superposition eye
C	contrast
D	diameter of lens or ommatidial facet
d	diameter of a photoreceptor or rhabdom
$\Delta\phi$	inter-receptor or inter-ommatidial angle
$\Delta\rho$	acceptance angle (object space) of a photoreceptor or rhabdom
f	focal length (posterior nodal distance)
F-number	f/D
I, O	sizes of image and object
i_1, i_2	angles of incidence and refraction
k	absorption coefficient (of photopigment in a receptor)
L	length of photoreceptor
λ	wavelength of light
n	refractive index or number of photons
n_1, n_2	refractive indices
ν	frequency
ν_{co}	cut-off frequency of the optical system (usually D/λ)
ν_s	sampling frequency of receptor mosaic (usually $1/2\Delta\phi$)
P	power of an optical system $(1/f)$
r	radius of curvature or reflectance of a surface
S	sensitivity of an eye (Eqn 3.6)
s	separation of the centers of adjacent receptors
U, V	distances of object and image from nodal point
u, v	distances of object and image from principal plane (Fig. 3.5)

References

General reading

We include in this section a number of the most important books in the history of the comparative study of animal eyes from the late nineteenth century to the present. Others can be found in the 'Further reading' linked to specific chapters. In the present section the books are given in chronological order.

Grenacher, H. (1879). *Untersuchungen über das Sehorgan der Arthropoden, insbesondere der Spinnen, Insecten und Crustaceen.* Vanderhoeck und Ruprecht, Göttingen.
[Impressively accurate account of the microanatomy of arthropod eyes.]
Exner, S. (1891). *The physiology of the compound eyes of insects and crustaceans.* English edition (1988) translated and annotated by R.C. Hardie. Springer, Berlin.
[Classic account of compound eye function.]
Hesse, R. (1908). *Das Sehen der niederen Tiere.* Fischer, Jena.
[The culmination of many years of excellent anatomical studies of a wide range of eyes.]
Walls, G.L. (1942) *The vertebrate eye and its adaptive radiation.* Cranbrook Institute, Bloomington Hills. Reprinted (1967) Hafner, New York.
[Unequalled book on the eyes of all vertebrates.]
Rochon-Duvigneaud, A. (1943). *Les yeux et la vision des vertébrés.* Masson, Paris. et Cie.
[Comprehensive account of vertebrate eye anatomy and histology.]
Duke-Elder, S. (1958). *The eye in evolution.* Henry Kimpton, London.
[Comprehensive, beautifully illustrated account of eyes of all kinds. Sourcebook for the older literature.]
Wolken, J.J. (1971). *Invertebrate photoreceptors: a comparative analysis.* Academic Press, New York.
[Account of invertebrate eye structure at the electron microscope level. Complemented by the fine chapter by R.M. Eakin (1972) Structure of invertebrate

photoreceptors. In: *Handbook of sensory physiology*, Vol. VII/1 (ed. Dartnall, H.J.A.) pp. 625–84. Springer, Berlin.]

Two more recent compilations with a wide coverage are:
Sinclair, S. (1985) *How animals see*. Croom Helm, London.
[Notable for its impressive photographs of eyes.]
Wolken, J.J. (1995) *Light detectors, photoreceptors, and imaging systems in nature.* Oxford University Press, Oxford.
[Wide-ranging book covering subjects from phototaxis to bio-mimetic engineering].

Further reading for each chapter

The material in this section is intended to complement and extend the material in the chapter itself.

Chapter 1

Conway-Morris, S. (1998). *The crucible of creation*. Oxford University Press, Oxford.
[Account of the Burgess Shale fauna, by a geologist who studied it.]
Goldsmith, T.H. (1990). Optimization, constraint and history in the evolution of eyes. *Quarterly Review of Biology* **65**, 281–322.
[A thoughtful account of the issues in eye evolution.]
Land, M.F. and Fernald R.D. (1992). The evolution of eyes. *Annual Review of Neuroscience* **15**, 1–29.
Nilsson, D.-E. (1996). Eye ancestry – old genes for new eyes. *Current Biology* **6**, 39–42.
[Brief discussion of molecular genetic evidence for early eye evolution.]

Chapter 2

Browman, H.I., Hawryshyn, C.W. (eds) (2001) Biology of ultraviolet and polarization vision. *Journal of Experimental Biology* **204**, 2383–2596.
Hecht, E., Zajac, A. (1997). *Optics* (3rd edn). Addison-Wesley, Reading Mass.
[A good all-purpose optics text.]
Lythgoe, J.N. (1979). *The ecology of vision*. Clarendon Press, Oxford.
[Aspects of vision in different environments.]
Mollon, J.D. and Sharpe, L.T. (eds) (1983). *Colour vision*: physiology and psychophysics. Academic Press, London.
Rodieck, R.W. (1998). *The first steps in seeing*. Sinauer, Sunderland Mass.
[Particularly good account of the importance of photons for vision.]

Wandell, B.A. (1995). *Foundations of vision*. Sinauer, Sunderland Mass. [Chapter 9 on colour vision is particularly useful.]

Chapter 3

Charman, (1991). The vertebrate dioptric apparatus. In: *Vision and visual dysfunction*, Vol. 2 (eds Cronly-Dillon, J.R. and Gregory, R.L.) pp. 82–117. Macmillan, Basingstoke.
[Concise account of the optics of vertebrate eyes.]
Pirenne, M.H. (1967). *Vision and the eye*. Chapman & Hall, London.
[Classic book by one of the discoverers of the photon limit in vision.]
Snyder, A.W. (1979). Physics of vision in compound eyes. In: *Handbook of sensory physiology*, Vol. VII/6A (ed. Autrum, H.) pp. 225–313. Springer, Berlin.
[Thorough account of compound eye resolution and sensitivity.]
Warrant, E.J. (1999). Seeing better at night: life style, eye design and the optimal strategy of spatial and temporal summation. *Vision Research* **39**, 1611–30.
[Good discussion of the adaptions of eyes to dim conditions.]
Warrant, E.J. and McIntyre, P.D. (1993). Arthropod eye design and the physical limits to spatial resolving power. *Progress in Neurobiology* **40**, 413–61.
[The resolving power of compound eyes.]

Chapter 4

Messenger, J.B. (1991). Photoreception and vision in molluscs. In: *Vision and visual dysfunction*, Vol. 2 (eds Cronly-Dillon, J.R.E. and Gregory, R.L.). pp. 364–97. Macmillan, Basingstoke.
Nicol, J.A.C. (1989). *The eyes of fishes*, University Press, Oxford.
Walls, G.L. (1967). *The vertebrate eye and its adaptive radiation*, Haffner, New York. (Reprint of 1942 edition published by the Cranbrook Institute of Science, Bloomington Hills, Michigan.)
[The classic reference book on vision in vertebrates.]

Chapter 5

Charman, W.N. (1991). [as Chapter 3: compact but wide-ranging account of vertebrate optics.]
Hughes, A. (1979). The topography of vision in mammals of contrasting life style: comparative optics and retinal organization. In: *Handbook of sensory physiology*, Vol. VII/5 (ed F. Crescitelli), pp. 613–756. Springer, Berlin.
Oyster, C. (2000). *The human eye*, Sinauer, Sunderland Mass.
[A good modern text on the human eye, with a comparative introduction.]
Walls, G.L. (1967). [as Chapter 4]

Chapter 6

Denton, E.J. (1970). On the organization of reflecting surfaces in some marine animals. *Philosophical Transactions of the Royal Society of London B* **258**, 285–313.

Fox, H.M. and Vevers, G. (1960). *The nature of animal colours*. Sidgwick and Jackson, London.

[Old, but still very useful.]

Herring, P.J. (1994). Reflective systems on aquatic animals. *Comparative Biochemistry & Physiology* **109A**, 513–46.

[Modern account of structure and function.]

Land, M.F. (2000). Eyes with mirror optics. *Journal of Optics A*: Pure & Applied Optics **2**, R44–R50.

[Recent review.]

Chapters 7 and 8

Exner, S. (1891). *The physiology of the compound eyes of insects and crustaceans*, English edition (1988) translated and annotated by R.C. Hardie. Springer, Berlin.

[Classic account of compound eye function.]

Land, M.F. (1999). Compound eye structure: matching eye to environment. In: *Adaptive mechanisms in the ecology of vision* (eds Archer, S.N. *et al.*), pp. 51–71. Kluwer, Dordrecht.

[An ecological account of compound eye vision.]

Snyder, A.W. (1979). Physics of vision in compound eyes. In: *Handbook of sensory physiology*. Vol. VII/6A (ed. Autrum, H.), pp. 225–313. Springer, Berlin.

[Tough, but thorough.]

Stavenga, D.G. and Hardie, R.C. (eds) (1989). *Facets of vision*. Springer, Berlin.

[Articles by most of the authors active in the field at the time.]

Wehner, R. (1981). Spatial vision in arthropods. In: *Handbook of sensory physiology*, Vol. VII/6C (ed. Autrum, H.), pp. 287–616. Springer, Berlin.

[Encyclopaedic account of behaviour and its relation to visual function.]

Chapter 9

Carpenter, R.H.S. (1988). *Movements of the eyes* (2nd edn) Pion, London.

[Excellent readable textbook on human eye movements.]

Land, M.F. (1999). Motion and vision: why animals move their eyes. *Journal of Comparative Physiology A* **185**, 341–52.

[Recent review of eye movements across the animal kingdom.]

Srinivasan, M.V. and Venkatesh, S. (eds) (1997). *From living eyes to seeing machines*. Oxford University Press, Oxford.

[Contains a number of useful articles on 'active' vision.]

Walls, G.L. (1962). The evolutionary history of eye movements. *Vision Research* **2**, 69–80.
[Early wisdom on the functions of eye movements in animals.]
Yarbus, A. (1967). *Movements of the eyes*. Plenum Press, New York.
[Classic book on how we use our eyes.]

References

This section contains books and papers quoted in the text. These have been selected from the vast potential pool of literature as being either seminal papers (with a bias to the more recent), useful reviews, or papers from which our figures were derived (as cited in the figure captions).

Autrum, H. (1981). Light and dark adaptation in invertebrates. In: *Handbook of sensory physiology*, Vol. VII/6C (ed. Autrum, H.), pp. 1–91. Springer, Berlin.

Baccetti, B. and Bedini, C. (1964). Research on the structure and physiology of the eyes of a lycosid spider I. – Microscopic and ultramicroscopic structure. *Archives Italiennes de Biologie* **102**, 97–122.

Bernard, G.D. and Miller, W.H. (1968). Interference filters in the corneas of Diptera. *Investigative Ophthalmology.* **7**, 416–434.

Blest, A.D. (1985). The fine structure of spider photoreceptors in relation to function. In: *Neurobiology of arachnids* (ed. Barth, F.G.), pp. 79–102. Springer, Berlin.

Briggs, D.E.G. (1991). Extraordinary fossils. *American Scientist*, **79**, 130–41.

Brocco, S.L. and Cloney, R.A. (1980). Reflector cells in the skin of *Octopus dofleini*. *Cell & Tissue Research* **205**, 167–86.

Bryceson, K.P. and McIntyre, P. (1983). Image quality and acceptance angle in a reflecting superposition eye. *Journal of Comparative Physiology* **151**, 367–80.

Buschbeck, E., Ehmer, B., Hoy, R. (1999). Chunk versus point sampling: visual imaging in a small insect. *Science* **286**, 1178–80.

Carpenter, R.H.S. (1988). *Movements of the eyes* (2nd edn.) Pion, London.

Charman, W.N. (1991). The vertebrate dioptric apparatus. In: *Vision and visual dysfunction*. Vol. 2. (eds Cronly-Dillon, J.R. and Gregory, R.L.), pp. 82–117. Macmillan, Basingstoke.

Chaudhuri, A., Hallett, P.E. and Parker, J.A. (1983). Aspheric curvatures, refractive indices and chromatic aberration for the rat eye. *Vision Research* **23**, 1351–64.

Chittka, L. (1996). Does bee color vision predate the evolution of flower color? *Naturwissenschaften* **83**, 136–38.

Collett, T.S. and Harkness, L. (1982). Depth vision in animals. In: *Analysis of visual behavior* (eds Ingle, D.J., Goodale, M.A., Mansfield, R.J.), pp. 111–76. MIT Press, Cambridge Mass.

Collett, T.S. and Land, M.F. (1975). Visual control of flight behaviour in the hoverfly *Syritta pipiens* L. *Journal of Comparative Physiology* **99**, 1–66.

Collett, T.S., and Lehrer, M. (1993). Looking and learning: a spatial pattern in the orientation flight of the wasp *Vespula vulgaris*. Proceedings of the Royal Society of London B **252**, 129–34.

Collin, S.P., and Pettigrew, J.D. (1988). Retinal topography in reef teleosts. I & II. *Brain Behavior & Evolution* **31**, 269–95.

Collin, S.P., Hoskins, R.V., and Partridge, J.C. (1998). Seven retinal specializations in the tubular eye of the pearleye, *Scopelarchus michaelsarsi*: a case study in visual optimization. *Brain Behavior & Evolution* **51**, 291–314.

Conway-Morris, S. (1998). *The crucible of creation*. Oxford University Press, Oxford.

Conway-Morris, S. and Whitington, H.B. (1985). Fossils of the Burgess Shale. A national treasure in Yoho National Park, British Columbia. *Geological Survey of Canada, Miscellaneous Reports*, **43**, 1–31.

Dacke, M., Nilsson, D.-E., Warrant, E.J., Blest, A.D., Land, M.F., and O'Carroll, D.C. (1999). A new compass organ in spiders, using built-in polarizers. *Nature* **401**, 470–2.

Dahmen, H. (1991) Eye specialization in water striders: an adaptation to life in a flat world. *Journal of Comparative Physiology A* **169**, 623–32.

Denton, E.J. (1970). On the organization of reflecting surfaces in some marine animals. *Philosophical Transactions of the Royal Society of London B* **258**, 285–313.

Denton, E.J. (1990). Light and vision at depths greater than 200m. In: *Light and life in the sea* (eds Herring, P.J., Campbell, A.K., Whitfield, M. and Maddock, L.), pp. 127–48. Cambridge University Press.

Denton, E.J. and Nicol, J.A.C. (1965). Reflexion of light by external surfaces of the herring, *Clupea harengus*. *Journal of the Marine Biological Association of the UK* **45**, 711–38.

Easter, S.S., Johns, P.R., and Heckenlively, D. (1974). Horizontal compensatory eye movements in goldfish (*Carrassius auratus*). I. The normal animal. *Journal of Comparative Physiology* **92**, 23–35.

Eisner, T., Silberglied, R.E., Aneshansley, D., Carrel, J.E., and Howland, H.C. (1969) Ultraviolet video-viewing: the television camera as an insect eye. *Science* **166**, 1172–74.

Exner, S. (1891). *The physiology of the compound eyes of insects and crustaceans*. Translated from the German by R. C. Hardie (1989), republished by Springer-Verlag.

Forster, L. (1985). Target discrimination in jumping spiders. In: *Neurobiology of arachnids* (ed. Barth, F.G.), pp. 249–74. Springer, Berlin.

Fox, D.L. (1953). *Animal biochromes and structural colours*. University Press, Cambridge.

Fox, H.M. and Vevers, G. (1960). *The nature of animal colours*. Sidgwick & Jackson, London.

Franceschini, N. (1975). Sampling of the visual environment by the compound eye of the fly: fundamentals and applications. In: *Photoreceptor optics* (eds Snyder, A.W. and Menzel, R.), pp. 98–125. Springer, Berlin.

Fraenkel, G.S., and Gunn, D.L. (1961). *The orientation of animals*. Dover, New York.

Friederichs, H.F. (1931) Beiträge zur Morphologie und Physiologie der Sehorgane der Cicindeliden (Col.). *Zeitschrift für Morphologie und Ökologie der Tiere* **21**, 1–172.

Gehring, W.J., and Ikeo, K. (1999). Pax-6: mastering eye morphogenesis and eye evolution. *Trends in Genetics* 15, 371–7.

Gibson, J.J. (1979). *The ecological approach to visual perception*. Houghton-Mifflin, Boston.

Gilchrist, I.D., Brown, V., and Findlay, J.M. (1997). Saccades without eye movements. *Nature* 390, 130–1.

Goodman, L.J. (1981). Organization and physiology of the insect dorsal ocellar system. In: *Handbook of sensory physiology*. Vol. VII/6C (ed Autrum, H.), pp. 201–86. Springer, Berlin.

Goodale, M.A., Ellard, C.G., and Booth, L. (1990). The role of image size and retinal motion in the computation of absolute distance by the mongolian gerbil (*Meriones unguiculatus*). *Vision Research* **30**, 399–413.

Görner, P., and Claas, B. (1985). Homing behavior and orientation in the funnel-web spider, *Agalena labyrinthica* Clerk. In: *Neurobiology of arachnids* (ed Barth, F.G.), pp. 275–97. Springer, Berlin.

Gould, S.J. (1989). *Wonderful life. The Burgess Shale and the nature of history*. p. 347. Norton, New York.

Gregory, R.L. (1991). Origins of eyes – with speculations on scanning eyes. *Vision and visual dysfunction*, Vol. 2. (eds Cronly-Dillon, J.R., and Gregory, R.L.), pp. 52–9. Macmillan, Basingstoke.

Greuet, C. (1982). Photorécepteurs et phototaxie des flagellés et des stades unicellulaires d'organismes inférieures. *Annales de Biologie* **21**(2), 98–141.

Hardy, A. (1956). *The open sea*. Collins, London.

Hateren, J. van (1989). Photoreceptor optics, theory and practice. In: *Facets of vision* (eds Stavenga, D.G., and Hardie, R.C.), pp. 74–89. Springer, Berlin.

Hateren, J.H. van, Hardie, R.C., Rudolph, A., Laughlin, S.B., and Stavenga, D.G. (1989). The bright zone, a specialised dorsal eye region in the male blowfly *Chrysomyia megalocephala*. *Journal of Comparative Physiology A* **164**, 297–308.

Herring, P.J. (1994). Reflective systems in aquatic animals. *Comparative Biochemistry & Physiology* **109A**, 513–46.

Hesse, R. (1908). *Das Sehen der niederen Tiere*. Fischer, Jena.

Horridge, G.A. (1978). The separation of visual axes in apposition compound eyes. *Philosophical Transactions of the Royal Society of London B* **285**, 1–59.

Howard, J., Dubs, A., and Payne, R. (1984). The dynamics of photo-transduction in insects. A comparative study. *Journal of Comparative Physiology A* **154**, 707–18.

Hughes, A. (1977). The topography of vision in mammals of contrasting life style: comparative optics and retinal organization. In: *Handbook of sensory physiology*, Vol. VII/5 (ed Crescitelli, F.), pp. 613–756. Springer, Berlin.

Imms, A.D. (1956), *Insect natural history* (2nd edn). Collins, London.

Jagger, W.S. (1992). The optics of the spherical fish lens. *Vision Research* **32**, 1271–84.

Kirschfeld, K. (1976). The resolution of lens and compound eyes. In: *Neural principles in vision* (eds Zettler, F. and Weiler, R.), pp. 354–370. Springer, Berlin.

Kröger, R.H.H., Campbell, M.C.W., Fernald, R.D., and Wagner, H.-J. (1999). Multifocal lenses compensate for chromatic defocus in vertebrate eyes. *Journal of Comparative Physiology A* **184**, 361–369.

Kuchiiwa, T., Kuchiiwa, S. and Teshirogi, W. (1991). Comparative morphological studies on the visual systems in a binocular and a multi-ocular species of freshwater planarian. *Hydrobiologia*, **227**, 241–249.

Kunze, P. (1979). Apposition and superposition eyes. In: *Handbook of sensory physiology*, Vol. VII/6A (ed. Autrum, H.) pp. 442–502. Springer, Berlin.

Labhart, T., and Nilsson, D.-E. (1995). The dorsal eye of the dragonfly *Sympetrum*: specializations for prey detection against the blue sky. *Journal of Comparative Physiology A* **176**, 437–53.

Land, M.F. (1965). Image formation by a concave reflector in the eye of the scallop, *Pecten maximus*. *Journal of Physiology* (London) **179**, 138–53.

Land, M.F. (1969). Movements of the retinae of jumping spiders in response to visual stimuli. *Journal of Experimental Biology* **51**, 471–93.

Land, M.F. (1972). The physics and biology of animal reflectors. *Progress in Biophysics & Molecular Biology* 24, 77–106.

Land, M.F. (1980). Eye movements and the mechanism of vertical steering in euphausiid Crustacea. *Journal of Comparative Physiology* **137**, 255–65.

Land, M.F. (1981). Optics and vision in invertebrates. In: *Handbook of sensory physiology*, Vol. VII/6B (ed. Autrum, H.) pp. 471–592. Springer, Berlin.

Land, M.F. (1982). Scanning eye movements in a heteropod mollusc. *Journal of Experimental Biology* **96**, 427–30.

Land, M.F. (1984). The resolving power of diurnal superposition eyes measured with an ophthalmoscope. *Journal of Comparative Physiology A* **154**, 515–33.

Land, M.F. (1984). Crustacea. In: *Photoreception and vision in invertebrates* (ed. Ali, M.A.) pp. 401–38. Plenum Press, New York.

Land, M.F. (1985). The morphology and optics of spider eyes. In: *Neurobiology of arachnids* (ed Barth, F.G.) pp. 53–78. Springer, Berlin.

Land, M.F. (1989). Variations in the structure and design of compound eyes. In: *Facets of vision* (eds Stavenga, D.G., and Hardie, R.C.) pp. 90–111. Springer, Berlin.

Land, M.F. (1997). Visual acuity in insects. *Annual Review of Entomology* **42**, 147–77.

Land, M.F. (1999). Motion and vision: why animals move their eyes. *Journal of Comparative Physiology A* **185**, 341–52.

Land, M.F. (1999). Compound eye structure: matching eye to environment. In: *Adaptive mechanisms in the ecology of vision* (eds. Archer, S.N. et al.) pp. 51–71. Kluwer, Dordrecht.

Land, M.F., and Fernald, R.D. (1992). The evolution of eyes. *Annual Review of Neuroscience* **15**, 1–29.

Land, M.F. and Nilsson, D.-E. (1990). Observations on the compound eyes of the deep-sea ostracod *Macrocypridina castanea*. *Journal of Experimental Biology* **148**, 221–33.

Land, M.F. and Osorio, D. (1990). Waveguide modes and pupil action in the eyes of butterflies. *Proceedings of the Royal Society of London B* **241**, 93–100.

Land, M.F., Marshall, N.J., Brownless, D., Cronin, T. (1990). The eye movements of the mantis shrimp *Odontodactylus scyllarus* (Crustacea: Stomatopoda). *Journal of Comparative Physiology A* **167**, 155–66.

Land, M.F., Mennie, N., and Rusted, J. (1999), The roles of vision and eye movements in the control of activities of daily living. *Perception* **28**, 1311–28.

Layne, J.E. (1998). Retinal location is the key to identifying predators in fiddler crabs (*Uca pugilator*). *Journal of Experimental Biology* **201**, 2253–61.

Layne, J.E., Wicklein, M., Dodge, F.A., and Barlow, R.B. (1997). Prediction of maximum allowable retinal slip in the fiddler crab, *Uca pugilator*. *Biological Bulletin of Woods Hole* **193**, 202–03.

Lehrer, M., Srinivasan, M.V., Zhang, S.W., and Horridge, G.A. (1988). Motion cues provide the bee's visual world with a third dimension. *Nature* **332**, 356–57.

Levi-Setti, R. (1993). *Trilobites*. Chicago University Press, Chicago.

Lockett, N.A. (1977). Adaptations to the deep-sea environment. In: *Handbook of sensory physiology*, Vol. VII/5 (ed. Crescitelli, F.) pp. 67–192. Springer, Berlin.

Lythgoe, J.N. (1979). *The ecology of vision*. Clarendon Press, Oxford.

Lythgoe, J.N., and Shand, J. (1989). The structural basis for iridescent colour changes in dermal and corneal iridophores in fish. *Journal of Experimental Biology* **141**, 313–25.

McIntyre, P., and Caveney, S. (1998). Superposition optics and the time of flight of onitine dung beetles. *Journal of Comparative Physiology A* **183**, 45–60.

Mallock, A. (1894). Insect sight and the defining power of composite eyes. *Proceedings of the Royal Society of London B* **55**, 85–90.

Marshall, N.B. (1979). *Developments in deep-sea biology*. Blandford, Poole.

Marshall, N.J., Oberwinkler, J. (1999). The colourful world of the mantis shrimp. *Nature* **401**, 873–74.

Marshall, J., Cronin, T.W., Shashar, N., and Land, M. (1999). Behavioural evidence for polarization vision in stomatopods reveals a potential channel for communication. *Current Biology* **9**, 755–58.

Martin, G.R. (1983). Schematic eye models in vertebrates. *Progress in Sensory Physiology* **4**, 43–81.

Martin, G.R. (1985). Eye. In: *Form and function in birds*, Vol. 3, pp. 311–73. Academic, London.

Messenger, J.B. (1981). Comparative physiology of vision in molluscs. In: *Handbook of sensory physiology*, Vol. VII/6C (ed Autrum, H.), pp. 93–200. Springer, Berlin.

Menzel, R. (1979). Spectral sensitivity and color vision in invertebrates. In: *Handbook of sensory physiology*, Vol. VII/6A (ed. Autrum, H.), pp. 503–80. Springer, Berlin.

Miller, W.H., and Bernard, G.D. (1968). Butterfly glow. *Journal of Ultrastructural Research* **24**, 286–94.

Millodot, M. and Sivak, J.G. (1979). Contribution of the cornea and lens to the spherical aberration of the eye. *Vision Research* **19**, 685–87.

Montani, R., Rothschild, B.M., and Wahl, W.Jr. (1999). Large eyeballs in diving ichthyosaurs. *Nature* **402**, 747.

Munk, O. (1970). On the occurrence and significance of horizontal and band-shaped retinal areae in teleosts. *Videnskabelige Meddelelser tra Dansk Naturhistorisk Forening* **133**, 85–120.

Muntz, W.R.A., and Raj, U. (1984). On the visual system of *Nautilus pompilius*. *Journal of Experimental Biology* **109**, 253–63.

Newell, G.E. (1965). The eye of *Littorina littorea*. *Proceedings of the Zoological Society of London* **144**: 75–86

Nicol, J.A.C. (1989). *The eyes of fishes*. Oxford University Press, Oxford.

Nicol, J.A.C., and Arnott, H.J., and Best, A.C.G. (1973). Tapeta lucida in bony fishes. *Canadian Journal of Zoology* **51**, 69–81.

Nilsson, D.-E. (1988). A new type of imaging optics in compound eyes. *Nature* **332**, 76–8.

Nilsson, D.-E. (1989). Optics and evolution of the compound eye. In: *Facets of vision* (eds Stavenga, D.G., and Hardie, R.C.), pp. 30–73. Springer, Berlin.

Nilsson, D.-E. (1990). Three unexpected cases of refracting superposition eyes in crustaceans. *Journal of Comparative Physiology A* **167**, 71–8.

Nilsson, D.-E. (1994). Eyes as optical alarm systems in fan worms and ark clams. *Philosophical Transactions of the Royal Society of London B* **346**, 195–212

Nilsson, D.-E. (1996). Eye ancestry – old genes for new eyes. *Current Biology* **6**, 39–42.

Nilsson, D.-E. and Pelger, S. (1994). A pessimistic estimate of the time required for an eye to evolve. *Proceedings of the Royal Society of London B* **256**, 53–8.

Nilsson, D.-E. and Modlin, R.F. (1994). A mysid shrimp carrying a pair of binoculars. *Journal of Experimental Biology* **189**, 213–36.

Nilsson, D.-E. and Ro, A.-I. (1994). Did neural pooling for night vision lead to the evolution of neural superposition eyes? *Journal of Comparative Physiology A* **175**, 289–302.

Nilsson, D.-E., Hamdorf, K., and Höglund, G. (1992). Localization of the pupil trigger in insect superposition eyes. *Journal of Comparative Physiology A* **170**, 217–26.

Nilsson, D.-E., Land, M.F. and Howard, J. (1988). Optics of the butterfly eye. *Journal of Comparative Physiology A* **162**, 341–66.

Ott, M. and Schaeffel, F. (1996). A negatively powered lens in the chamaeleon. *Nature* **373**, 692–94.

Packard, A. (1972). Cephalopods and fish: the limits of convergence. *Biological Reviews* **47**, 241–307.

Parker, A.R. (2000). 515 million years of structural colour. *Journal of Optics A: Pure and Applied Optics* **2**, R15–R28.

Partridge, J.C., and Douglas, R.H. (1995). Far-red sensitivity of dragon fish. *Nature* **375**, 21–22.

Paul, H., Nalbach, H.-O., Varjú, D. (1990). Eye movements in the rock crab *Pachygrapsus marmoratus* walking along straight and curved paths. *Journal of Experimental Biology* **154**, 81–97.

Paulus, H.F. (1979). Eye structure and the monophyly of the arthropoda. In: *Arthropod phylogeny* (ed. Gupta, A.P.), pp. 299–383. Van Nostrand Reinhold, New York.

Pedler, C. (1963). The fine structure of the tapetum cellulosum. *Experimental Eye Research* **2**, 189–95.

Pettigrew, J.D., Collin, S.P., and Ott, M. (1999). Convergence of specialised behaviour, eye movements and visual optics in the sandlance (Teleostei) and the chameleon (Reptilia). *Current Biology* **9**, 421–4.

Pix, W., Zanker, J.M., and Zeil, J. (2000). The optomotor response and spatial resolution in the male *Xenox vesparum* (Strepsiptera). *Journal of Experimental Biology* **203**, 3397–409.

Pumphrey, R.J. (1961). Concerning vision. In: *The cell and the organism* (eds Ramsay, J.A., and Wigglesworth, V.B.), pp. 193–208. Cambridge University Press, Cambridge.

Purnell, M.A. (1995). Large eyes and vision in conodonts. *Lethaia*, **28**, 187–8.

Reymond, L. (1985). Spatial visual acuity of the eagle *Aquila audax*: a behavioural, optical and anatomical investigation. *Vision Research* **25**, 1477–91.

Rhode, K and Watson, N.A. (1991) Ultrastructure of pigmented photoreceptors of larval *Multicotyle purvisi* (Trematoda, Aspidogastrea). *Parasitological Research* **77**: 485–90.

Rossel, S. (1979). Regional differences in photoreceptor performance in the eye of the praying mantis. *Journal of Comparative Physiology A* **131**, 95–112.

Rossel, S. (1980). Foveal fixation and tracking in the praying mantis. *Journal of Comparative Physiology A* **139**, 307–31.

Rossel, S. (1989). Polarization sensitivity in compound eyes. In : *Facets of vision* (eds Stavenga, D.G., and Hardie, R.C.), pp. 298–316. Springer, Berlin.

Salvini-Plawen, L. von, and Mayr, E. (1977). On the evolution of photoreceptors and eyes. *Evolutionary Biology* **10**, 207–63.

Schaeffel, F., Glasser, A., and Howland, H.C. (1988). Accommodation, refractive error and eye growth in chickens. *Vision Research* **28**, 639–57.

Scherer, C., and Kolb, G. (1987). Behavioral experiments on the visual processing of color stimuli in *Pieris brassicae* L. (Lepidoptera). *Journal of Comparative Physiology A* **160**, 645–56.

Schilstra, C., and Hateren, J.H. (1988). Stabilizing gaze in flying blowflies. *Nature* **395**, 654.

Schmitz, H., and Bleckmann, H. (1998). The photomechanic infrared receptor for the detection of forest fires in the beetle *Melanophila acuminata*. *Journal of Comparative Physiology A* **182**, 647–57.

Schwind, R. (1980). Geometrical optics of the *Notonecta* eye: adaptations to optical environment and way of life. *Journal of Comparative Physiology* **140**, 59–69.

Schwind, R. (1983). A polarization-sensitive response of the flying water bug *Notonecta glauca* to UV light. *Journal of Comparative Physiology A* **150**, 87–91.

Sivak, J.G. (1976). The accommodative significance of the 'ramp' retina in the eye of the stingray. *Vision Research* **16**, 945–50.

Sivak, J.G., Hildebrand, T., and Lebert, C. (1985). Magnitude and rate of accommodation in diving and non-diving birds. *Vision Research* **25**, 925–33.

Snyder, A.W. (1979). Physics of vision in compound eyes. In: *Handbook of sensory physiology*, Vol. VII/6A (ed. Autrum, H.), pp. 225–313. Springer, Berlin.

Snyder, A.W. and Miller, W.H. (1978). Telephoto lens system of falconiform eyes. *Nature* **275**, 127–9.

Sobel, E.C. (1990). The locust's use of motion parallax to measure distance. *Journal of Comparative Physiology A* **167**, 579–88.

Srinivasan, M.V., Lehrer, M., Zhang, S.W., and Horridge, G.A. (1989). How honeybees measure their distance from objects of unknown size. *Journal of Comparative Physiology A* **165**, 605–13.

Stange, G. (1981). The ocellar component of flight equilibrium control in dragonflies. *Journal of Comparative Physiology A* **141**, 335–47.

Stavenga, D.G. (1979). Pseudopupils of compound eyes. In: *Handbook of sensory physiology*, Vol. VII/6A (ed. Autrum, H.), pp. 225–313. Springer, Berlin.

Steinbrecht, R.A., Mohren, W., Pulker, H.K. and Schneider, D. (1985). Cuticular interference reflectors in the golden pupae of danaine butterflies. *Proceedings of the Royal Society of London B* **226**, 367–90.

Timney, B. and Keil, K. (1992). Visual acuity in the horse. *Vision Research* **32**, 2289–93.

Vogt, K. (1980). Die Spiegeloptik des Flusskrebsauges. The optical system of the crayfish eye. *Journal of Comparative Physiology* **135**, 1–19.

Walls, G.L. (1942). *The vertebrate eye and its adaptive radiation*. Cranbrook Institute, Bloomington Hills. Reprinted (1967) Hafner, New York.

Walls, G.L. (1962). The evolutionary history of eye movements. *Vision Research* **2**, 69–80.

Warrant, E.J. (1999). Seeing better at night: life style, eye design and the optimum strategy of spatial and temporal summation. *Vision Research* **39**, 1611–30.

Warrant, E.J., and McIntyre, P.D. (1990). Screening pigment, aperture and sensitivity in the dung beetle superposition eye. *Journal of Comparative Physiology A* **167**, 805–15.

Warrant, E.J., and Nilsson, D.-E. (1995). The absorption of white light by photo-receptors. *Vision Research* **38**, 195–207.

Warrant, E., Bartsch, K., and Günther, C. (1999). Physiological optics in the hummingbird hawkmoth: a compound eye without ommatidia. *Journal of Experimental Biology* **202**, 497–511.

Warrant, E., Porombka, T., and Kirchner, W.H. (1996). Neural image enhancement allows honeybees to see at night. *Proceedings of the Royal Society of London B* **263**, 1521–26.

Wehner, R. (1981). Spatial vision in arthropods. In: *Handbook of sensory physiology*, Vol. VII/6C (ed. Autrum, H.), pp. 287–616. Springer, Berlin.

Wehner, R. (1987). 'Matched filters' – neural models of the external world. *Journal of Comparative Physiology A* **161**, 511–31.

Williams, D.S. and McIntyre, P. (1980). The principal eyes of a jumping spider have a telephoto component. *Nature* **288**, 578–80.

Xianguang, H. and Bergström, J (1997) Arthropods of the Lower Cambrian Chengjiang fauna, southwest China. *Fossils & Strata* **45**: 1–116.

Yarbus, A. (1967). *Movements of the eyes*. Plenum Press, New York.

Young, J.Z. (1964). *A model of the brain*. Oxford University Press, Oxford.

Zeil, J. (1983). Sexual dimorphism in the visual system of flies: the free flight behaviour of male Bibionidae (Diptera). *Journal of Comparative Physiology* **150**, 395–412.

Zeil, J. (1993). Orientation flights of solitary wasps (*Cerceris*; Sphecidae; Hymenoptera). *Journal of Comparative Physiology A* **172**, 189–222.

Zeil, J., Nalbach, G., and Nalbach, H.-O. (1989). Spatial vision in a flat world: optical and neural adaptations in arthropods. In: *Neurobiology of sensory systems* (eds Singh, R.N., and Strausfeld, N.J.), pp. 123–37. Plenum Press, New York.

Index

Abalone, eye 56–57
Acceptance angle 130–1, **134**, 188–9
Accommodation
    lens function 85–7
    reptiles and birds 86
    vertebrate mechanisms **85**
Activity, day and night 84
Adaptation
    dark, superposition eyes **165**
    light and dark
        eye-glow monitoring 165
        mechanisms 138–**139**
        superposition eyes 164–6
Aeschna multicolor, eye **148**
Airy
    diffraction image 162
    diffraction pattern **134**
        origins 40–1
    disc 41
        diameter 163
Amegilla, pseudopupil **140**
Amphibious eyes **93–4**
Amphioxus 3
Amphitretus pelagicus 68
Anableps, accommodation **93–4**
Anartia sp, tapetum chitin-air multilayer **115**
Anax junius 150
Anchoa mitchilli, tapetum **115**
Anemone **Plate 1**
Angular velocities 198–**199**
Animal groups, evolutionary relationships **14**
Anomalocaris 125
Anomuran hermit crab 177
Ant-lion, larval ocelli **100**
Aplocheilus 153
Aplocheilus lineatus, visual streaks **67**

Apposition 127–31
    afocal 133
Apposition eye
    acuity distribution patterns 142
    ancestral 174
    compound 125–55
    ecological variations 142–53
    function **126**
    image formation mechanisms **132**–3
    optical comparison **129**
    structure **128**
Aptychotrema rostrata **Plate 1**
Aquatic eyes 56–71
Arca 105, 126–**127**
Arctosa variana 98
Aristostomias 69
Ark shells 105, 126
Artemia **139**
Arthropoda 126
Ascalaphus, pigment migration 166
Atalophlebid mayflies 177

Bay anchovy, tapetum **115**
Bdellocephala brunnea, receptor cells **4**
Bee
    corneal lens **132**
    distance learning **191**–2
    drone
        resolution distribution **147**
        sexual pursuit 149
    photoreceptor spectral sensitivity curves **25**, 29
Beetle, superposition eyes 157
Bibio marci 150
Bibionid fly
    male

eyes 150
    resolution distribution 147–**148**
Bioluminescence 21
Birds, eye movements 185
Black meadow experiment **191**–2
Blowfly
    male 149
        resolution distribution 146–**147**
    vision 102
Blur
    avoidance 193
    image degradation 200
    rule 189
Bragg, W.L. 19
Branchiomma 126–**127**
Burgess shale 1–2
Butterflies
    colours 118
    eye 168, **169**–72
        afocal 171
        magnification 170
        mode patterns 171
        optics **168**
        reflectors 118
    lens/lens-cylinder afocal combination **132**
    ommatidial receptive fields **144**
    optical system 133
    pupa, reflecting multilayer cuticles 123
    resolution pattern 167

Callinectes 137–**138**
Callionymus lyra, reflectors 118

Calliphora 146
   sexual pursuit 149
   vision 102
Cambrian explosion 1–2
Camouflage
   reflecting 121–3
      in sea **122**
Cardium 109
Cartesian convention 76
Cat, tapetum **115**
   ganglion cell pattern **92**
   pupil shape **89**
Catfish 89
   pupil shape 89
Centroptilum sp. **Plate 3**
Cephalopholis, ganglion cell pattern
   66–7, **Plate 1**
Cephalopods 63–8
Cerceris, orientation flights 198
Cerceris rybyensis, flight path **199**
Chalarus 149
Chamaeleon, eye structure **83**
Chamaeleo oshaugnessyi **82**
Chiasms 154
China, Chengjiang fauna 2
Chitin 116–17, 118, 119–**120**
Chromatic aberration 42, **43–4**, **61**
Chromophore **28**, 30
Chrysomyia megalocephala
   149
Cicindela, larval eyes **101–2**
Cirolana 137–**138**
Clam, pigment-pit eyes 7–8
Clupea harengus, scale **Plate 2**
Clydagnathus 3
Cockles 109
Cod **64**
Colour 24–7
   structural 121
Colour vision 28–30
   dodeca-chromatic 195
   pigments needed **27**
   system 195
Contrast 24
   transfer function **38–9**, **50**
Copilia **70–1**
Cornea
   insect eyes 99–102
   lens **13**, **132**
   optics 73–7
   shape and spherical correction
      87
Corner reflector 173–4, **175**
   principle **175**

Cougar, eye structure **83**
Counter-illumination 122–3
Crab spiders, vision **97**
Crabs
   eye movements **184**
   flat beach 152
   rocky upper shore 152
   Xanthid 177
Crayfish
   mirror box eyes **172**
   superposition eyes 158
Crustaceans, double eyes 150–1
Cupiennius salei **113**
Cuttlefish, polarization vision
   31
Cypselurus heterurus,
   accommodation 94
Cystisoma 151
   double eyes 166

Daphnia, colour vision 130
Dark
   see also Adaptation
   adaptation mechanisms
      138–**139**
Decapod shrimp, superposition
   eye **164**
Deilephila, pigment migration
   165–6
Diffraction 34
   apposition compound eyes
      135–7
   and image **40**
   limit 39
   pattern 41
Dilophus sp **148**
Dinoflagellates, single cell eye 15
Dinopidae 97
Dinopis **96–7**, 99
Dioptromysis 197
Dioptromysis paucispinosa, double
   eyes **166–7**
Diptera 154
   neural superposition eyes 133
Dog, eye structure **83**
Dolichopodid, colours 118
Dragonet, reflectors 118
Dragonfly
   eye size 137
   hunting 150
   resolution distribution **147–148**

Drassodes 99
Dromedary, eye structure **83**
Drosophila melanogaster
   anomalous eyes 154
   pseudopupil **140**
Dung beetle 162
   nocturnal, refracting
      superposition eye **156**

Ear, semi-circular canals 181–182
Einstein, Albert 18, 23
Elephant seal **82**
Empid fly **148**, 153
Eriococcus 102
Euphausiid
   double eyes 166
   refractive index gradient **160**
   superposition eyes 157
Euploea core, pupa, reflecting
   multilayer cuticles 123,
      Plate 2
Euroleon, larval ocelli **100**
Exner, Sigmund 157–**158**
Eye
   see also Amphibious eye;
      Apposition eye; Aquatic eye;
      Superposition eye
   apposition compound **13**,
      125–55
   basic compound **13**
   colour scanning **195**
   compound 125, **127**
      sizes **136**
   development 1–2
   diurnal and nocturnal,
      differences 54
   evolution 6–15
      course and pace 7–10
   focal length **36–7**
   grating resolution **36–7**
   mirrors in 104–24
   movements 178–206
      detection 189–90
   optical types **13**
   pit **13**
   purpose 4–6
   reflecting sunshades 113
   refracting superposition
      compound **13**
   scanning, values *197*
   sequential modifications 9

'simple' 125–**126**
  single chambered **13**
  single and double 166–8
  structure variations **83**
Eye-glow
  superposition eyes **164**
  superposition pupil 164
Eye-in-space movements 183

Fiddler crab 152
Film
  optical thickness 114, 116
  surface, vision from below **153**
Firefly
  corneal image **157**
  refractive index gradient **160**
Fish
  camouflage 121–3
  eye 63–5
    and environment 66–9
    size 66
    sunshade 113
  focusing mechanisms 65
  multilayer mirrors 116
  non-spherical lens 69–71
Fixation 179–**180**
Flat surfaces, resolution **143**
Flatworm, planarian **4**
Flight
  forward, pattern 143–6
  and vegetation, resolution **143**
Flow-field 190–3
  distance measurement 193
  locomotor 192
  velocity, on retina **190**
Flowers, spectral reflectances 25
Fly
  corneal lens **132**
  stalk-eyed, eye movements
    184–**185**
  vision 130–1
Focal length
  definition **75**
  equivalent 79
Focal powers 77
Focus 42–3
Four-eyed fish, accommodation
  93–4
Fovea 179–80, *184*
Fresnel's formula 116
Gadus **64**

Ganglion cells, distribution 91,
  92–3
Gaze
  movements **184**
  stabilisation 178, **180**
  stabilising reflexes 191
  stability 189–90
Gecko, Tokay, eye **82, 89**
Genes, eye development 10
Gennadas 177
Gerbil, head-bob 193
Gerris **146, 148**, 153
Ghost crab
  acuity band 152
  pseudopupil **140**, 142
Gigantocypris, reflector eyes 109,
  110–11
Glare 174
Goldfish, eye movements 183–**184**
Gratings 36–7
Greenland, Sirius Passet fauna 2
Guanine 116–17
Gullstrand model 78–**79**

Habronattus americanus, **Plate 4**
Haematopota pluvialis, **Plate 3**
Haliotis 56–**57**
Hawk
  resolution 90
  telephoto optics **91**
Helix 63
Heteronympha merope **144**–5,
  169–70, **Plate 3**
High-pass filter 190
Hilara sp **148**
Histioteuthis **62**, 68
Hollardops mesocristata 154
Holochroal eyes 154
Homoptera 149
Horse
  behavioural resolution 90
  eye size 83–4
  pupil shape **89**
Horsefly, cornea **115**
Horseshoe crab
  compound eyes 125
  cornea 132
Housefly, sexual pursuit 149
Hoverfly
  object detection 189
  tracking behaviour **186**–7
Human cornea 43–4

Human eye
  model 77–81
  movements, types and roles *184*
Human optic nerve 93
Human rods, absorption spectra **25**
Human visual information
  179–83
Hummingbird hawkmoth 167
Huygens-Fresnel scheme **17**, 18–19
Hymenoptera, scanning
  procedures 198–**199**
Hypericum **Plate 1**
Hyperiid amphipod, apposition
  eyes 150–**151**
Hybomitra lasiophthalma, cornea
  115
Hypsicomus, compound eyes **127**
Hyrax, pupil shape **89**

Illuminance 22–3
Image
  apposition and superposition
    157
  erect, superposition eyes 157
  eye-glow test 164
  formation
    apposition eye mechanisms
      132–3
    by curved cornea **73**–4
    effect of medium **72**
  motion effects 187–93
  and object, curved surface
    relations **74**
  quality in superposition eye **162**
  retinal 187
  size inverse to aperture **38**, 42
  superposition eye, ray paths **158**
Imagery, superposition, nature of
  156–9
Insects
  corneal eyes 99–102
  flight behaviour as eye
    movement 186–7
  larval ocelli **100**
Integration time 48
Inter-ommatidial angle 130
Interference microscope 158
Invertebrates
  corneal optics 94–102
  eye types 13, 15
Irradiance 22–3
Isia, larval ocelli **100**

Jellyfish
  box, eyes 15
  cubomedusan, spherical lens
    63
Junonia villida, pseudopupil **140**,
  142
Jumping spiders 95–**96**
  fields of view **96**
  pattern recognition **196**–8
  principal eyes **97**

Krill *see* Euphausiid

Labidocera **62**–3
  scanning eye 197–8
Leafhoppers 149
Leander, superposition eye **164**
Lens 42–4
  in accommodation 85–7
  aquatic **13**
  cephalopod 65
  cylindrical 176
  evolution 56–71
    gradual 58–9
  fish 65
  land 72–103
    optical surface 72–3
  multi-surface, water-bugs **132**
  optics 73–7
  paths of rays **59**
  with refractive index gradients
    60–3
  single, superposition eye 167
  spherical 59–60, **62**
  structure 56
Lens cylinder 132–3, 159–**160**
  Limulus **132**
  Phronima **132**
Lens-pad 68
Lens/cornea
  variations
    land vertebrates 81–93
    sizes and shapes 81–4
Lethrinus, retina ganglion pattern
  66–7
Light
  adaptation mechanisms
    138–**139**

distribution in Airy disc **39**
emittance 23
environmental **25**
intensity 19–21
  measurement 21–3
  photometric measurement
    system 21
  radiometric measurement
    system 21
intensity change 179
interference **17**
low level effect 34–**35**
nature of 16–19
polarization 16
polarized, reception **29**
propagation **17**
quantum theory 19
refraction **17**
spectrum **25**
wavefronts **17**
Limnichyes fasciatus, lenticle 71
Limpet 56–**57**
Limulus 125
  apposition eye **160**
  cornea 132
  graded index lenses 133
  lens-cylinder **132**
  refractive index gradient **160**
Littorina 57
Lobelia **Plate 1**
Locust
  dorsal ocelli **101**
  peering 192
  resolution distribution **146**
Luminance 20, **22**–4
Lycosid spider, eye, tapeta 113
Lycosidae 97–8
Lynx, cornea and lens **84**
  eye structure **83**

Macroglossum, eye-glow 164,
  Plate 3
Macroglossum stellatarum 167
Macropipus 176
Man, eye structure **83**
Mantis shrimp
  colour scanning 194, **195**–6
  colour vision 130
  polarization vision 31
Mating zones 146–52
Matthiessen gradient 61

Matthiessen lens 62
Matthiessen's ratio 60
Maxwell, James Clerk, light waves
  18
Mayflies, Atalophlebid 177
Melanitis leda 170
Melanophila, infra-red radiation
  detection 25
Merganser, accommodation **93**–4
Metaphidippus, scanning speed
  198
Microcebus murinus **82**
Microvillous receptors 30–1
Mirounga leonina **82**
Mirror boxes **173**, 174–**175**
Mirrors
  animal, physical optics 114–17
  in eyes 104–24
  multilayer **115**
    colour 117
    display use 118, 121
    reflectance **117**
    in reflecting camouflage
      121–3
    spectral reflectance 119–20
    reflecting sunshades 113
Mnierpes macrocephalus **93**–4
Moths
  eye-glow 164
  image quality 162
  refractive index gradient **160**
  superposition eyes 157
  wing-scale **115**
Motion blur 187, **188**–9
Mouse, eye structure **83**
Mouse lemur **82**
Movements of the eye, A Yarbus 179
Multicotyle purvisi, eye spot **5**
Multilayer interference 114
Myopia, lower field 86
Mysid shrimp 197

Nautilus
  optomotor response **58**
  pinhole eye 56–**57**
Nematoscelis atlantica, double eyes
  166
Nematoscelis boopis, double eyes
  166
Nematoscelis megalops, double eyes
  166

Neon tetra, reflectors 118
Newton, Isaac 16–17, 104–5
Newton's series, colours 114
Nodal point **36–7**, 75–6, 80
Notodromas, reflector eyes 109
Notonecta 130, 132
    resolution distribution **153**
Nucleus of optic tract **182**
Nystagmus 181, *184*

Octopus
    eye **64**, **Plate 1**
        development 10
        muscles 65
        multilayer mirrors 116
        reflecting cell **115**
Oculomotor nuclei 181–**182**
Ocypode, pseudopupil **140**, 142
Ocypode ceratophthalmus 152
Odontodactylus **Plate 4**
    colour scanning 194, **195–6**
    colour vision 130
Ogre-faced spider **96**
Ommatidium 130–1
    definition 128
Onitis aygulus 162
Onitis belial 162
Onitis westermanni **156**
Opossum, eye structure **83**
Opsins *see* Photopigment proteins
Optical cut-off frequency 37–9
Optics
    receptor **45**
    superposition eye **161–3**
Optokinetic reflex 181–**182**,
    183–*184*, 191
Ostracod, reflector eyes 109
Owl, eye size 83–**4**
Owl-fly, pigment migration 166
Oxygyrus
    prey detection 193–**194**
    spherical lens eye 62

Paracheirodon innesi, reflectors 118
Pardosa prativaga, **Plate 3**
Patella 56–**57**
Pattern recognition, jumping
    spider **196–8**
Pecten, reflector eyes 105, **Plate 1**

Pectunculus 105, 126
Pelargonium **Plate 1**
Perga, larval ocelli **100**–101
Periwinkle **57**
Phacops 154
Phalanoides tristifica, image quality
    162
Phase retardation, formula 119
Phidippus, pattern recognition
    196–8
Phidippus johnsoni 97, 99
Phoebus rurina, ultraviolet
    markings 26
Photometric unit **22**
Photons 34–5
    available numbers 50–1
    low numbers
        consequences 48, 51
        effect 47
    scattering and reflection 31
    statistics and contrast detection
        49
Photophores, luminescent 123
Photopigment **25**, 51
    ratios of stimulation 27
Photopigment proteins 10, **12**,
    28
Photoreceptors 5–6, 7
    action time 178–9
    ciliary **11**
    directional 6, 7
    microvillar **11**
    optics **45**–8
    signals 20
Photuris sp, corneal image 157
Phronima, lens-cylinder **132–3**
Phronima sedentaria, eye **151**–2
Phrosina, resolution distribution
    147
Phrosina semi-lunata, divided eye
    151
Pigment, migration 165–6
Pigment cup 9
pigeon
    eye structure **83**
    ganglion cell pattern **92**
Pikaya **3**
Pipunculid fly, female 149
Planck, Max 18, 23
Polarization 30–1
    natural **30**
    navigation aid 31
Polarization vision 31
Polyphemus 148–9

Pontella, ventral eye **69**
Portia fimbriata **96**–7
Portunus 176
Praying mantis, scanning
    movements **192**
Prey capture zones 146–52
Prey detection, sea snail 193–**194**
Procavia, pupil shape **89**
Protula, compound eyes **127**
Pseudopupil **195**
    antidromic 142
    apposition eyes 139, **140**–42
    explanation **141**
    orthodromic 142
Pterotrachea, spherical lens eye **62**
Pupil
    contractable 44
    diameter, effect on resolution
        88
    form and function 87–9
    shapes in vertebrates 88–**89**
    superposition 164
Pursuit behaviour, requirement
    143

Rabbit, ganglion cell pattern **92**
Radiance 21–**22**
Radiometric unit **22**
Ramp retina 86
Rat, ganglion cell pattern **92**
Reflectance
    formula 119
    spectral
        distribution of quarter-
            wave multilayers **120**
        multilayer mirrors
            119–20
Reflection **114**
    law of **105**
Reflectors *see* Mirrors
Refractive index **59**–60
    gradient 60–**61**, 87, 160
        fish and cephalopod
            65
Resolution 34, 36–*39*, 90–1
    apposition eye 133–5
    and contrast loss **50**
    distribution **146**
    and eye design 46–8
    limits **35**
    loss in motion 187

Resolution (*contd*)
  representation **145**
  and scanning speed 198
  superposition eye 161–3
Retina
  adaptation mechanisms 20
  angular velocity and
    distance/angle **143**
  'area centralis' 92
  development 10
  ganglion cells distribution **92**
  motion pattern 190–3
  narrow 193, 196
  one-dimensional, rotational
    scanning 193–200
  organization 91–2
  sampling frequency 36–7, 47,
    53
Rhabdom 118, 129–31, **134–5**
Rhabdomeres 130–1
Rhamphomyia tephraea 153
Rock-pool fish **93–4**
Rose-de Vries law 48, 49

Sabella, compound eyes **127**
Saccade and fixate strategy
  179–**180**, 183–6, **185**
Saccadic suppression 183
Salticidae 97–8
Sampling frequency 37
Sandlance, lenticle 71
Sapphirina **70**
Sawfly, larval ocelli **100**
Scallop
  eye **106**
    image formation **108**
    image-forming reflector
      105–9
    images **107**
    lens 107–8
Scanning
  colour, mantis shrimp 194,
    195–6
  rotational, by one-dimensional
    retina 193–200
  speed and resolution 198
Schistocerca gregaria, dorsal ocelli
  101
Schizochroal eyes 154–5
Schmidt corrector plate 108
Scopelarchus **68**

Scypholanceola, mirror eye
  110–11
Sea, deep, light 67–9
Sea snail
  prey detection 193–**194**
  spherical lens eye **62**
Seals, cornea and lens 84
Sensitivity 34, 48–53
  adaptation 137
  animal eyes *52*
  apposition eye 137–8
  calculation for apposition and
    superapposition eyes *163*
  definition 51
  increasing **51–3**
  range *52–3*
  superposition eye 161–3
Shrimp
  decapod 177
  mirror box eyes **172**
  ray paths **175**
Simuliid fly 150
Size 53–4
  apposition compound eyes
    135, **136–7**
  reasons 82–4
Skipper butterfly
  image quality 162
  refractive index gradient **160**
Snakes, infra-red wavelengths
  24–5
Snell's law **17**
Sparassidae 97
  eye, tapeta 113
Spatial frequency 37–**38**
Spatial summation 53
Spherical aberration 42, **43–4**, 59,
  60
Sphingid moth, pigment
  migration 165–6
Sphodromantis lineola, scanning
  movements **192**
Spiders
  eyes **95–9**
  web-spinning 99
Squalus acanthias, mirror eye **111**
Squid
  eye development 10
  Japanese firefly squid 65
  mid-water 68
  eyes **62**
Stabilisation reflexes 181–2
Steradian 22
Stomatopoda 194–6

Streetsia challengeri, cylindrical eye
  151
Strepsipterans, anomalous eyes
  153–5
Stylocheiron elongatum, double
  eyes 167
Stylocheiron maximum, double eyes
  166–7
Superposition
  neural 131
  parabolic **176–7**
  reflecting 172–6
    mechanism 173
    ray paths **173**
  refracting 159–68
    ray paths **173**
Superposition eye 125, 156–77
  3 types **159**
  double **166**
  neural, optical comparison
    129
  reflecting **172**
Swimming crab 176
Sympetrum sp 150
Syritta pipiens **148–9**
  tracking behaviour **186–7**

Tapetum lucidum 112–13
Tegenaria **95**
Telescopes 159, 161
Temporal summation 53
Tenodera australasiae 148
Thomisidae 97
Tiger beetle, larval eyes **101–2**
Trachynotus falcatus, reflecting
  123
Tracking 183–*184*
Trilobites, anomalous eyes 153–5

Uca pugilator 152
Ultraviolet 24, **26**
Urania ripheus
  colours 118
  wing-scale **115**, **Plate 2**

Velocity signal 192
Vergence movements *184*
Vertebrate eye, G Walls 180

Vertebrate rod, diagrammatic
 section **28**
Vespula 198
Vestibulo-ocular reflex **180**,
 181–**182**, 183–*184*,
 191
Vision
 motion 189–90, 192
 spatial 4–5, 10
 evolution 11–**12**
Visual information, human
 179–83
Visual streaks 92, 146

Walcott, Charles 1
Wasp, orientation flights
 198
Watasenia scintillans 65
Water, light polarization 31
Water-flea, colour vision
 130
Wavelength 24–30
 specific behaviours 29–30
White peacock butterfly,
 tapetum chitin-air multilayer
 115
Wolf spider, vision 97–8

Xanderella 2
Xenos peckii, anomalous eyes **154**

Young, Thomas, slit experiment
 17–18

Zizina labradus 169
Zone
 acute **195**
 acute horizontal 152–3
 clear 156–7